Team Microscope Endodontics

Kehn E. Yapp • Carlos Aznar Portoles

Team Microscope Endodontics

Carr's Guidelines to Best Practice

Springer

Kehn E. Yapp
East Melbourne, VIC, Australia

Carlos Aznar Portoles
Heemstede, The Netherlands

ISBN 978-3-031-70187-0 ISBN 978-3-031-70188-7 (eBook)
https://doi.org/10.1007/978-3-031-70188-7

This Springer imprint is published by the registered company Springer Nature Switzerland AG
The registered company address is: Gewerbestrasse 11, 6330 Cham, Switzerland

Foreword

With the introduction of this handbook, an important evolution in the development of microscopic endodontics has begun. Two thousand years ago, Aristotle maintained that the highest form of expertise (wisdom) consisted of possessing and combining theoretical understanding with practical knowledge. This handbook, and its associated demonstrations, validates that powerful insight by providing the theoretical basis for microscopic team excellence as well as the practical knowledge needed to master the required skills so necessary for proficient microscope use.

This combination of theoretical and practical knowledge has taken a long time coming to fruition in endodontics, and this handbook is a significant achievement in advancing that goal. Anyone who reads this handbook, and applies its lessons, improves their chance of reaching their full potential, both individually and as a team. They will also develop a more profound understanding of the role that office design and ergonomics play in both skill development and team performance.

I could not be more proud to have these two authors take these concepts and develop them systematically and rigorously. This handbook is an invaluable addition to our specialty and will help every single clinician who applies its methods.

San Diego, CA, USA — Gary B. Carr, Endodontist

Preface

The use of the dental operating microscope is one of the greatest advances in endodontics. The introduction of magnification with co-axial light has allowed clinicians to work in a more effective and predictable way, regardless of treatment complexity. Numerous new techniques and instruments have been developed from the use of the microscope that significantly improved the outcomes for both patients and clinicians over existing practices.

Anyone can develop the ability to use a microscope for all endodontic procedures. These skills are critical for not only the health and wellbeing of the clinical staff, but for ensuring the highest level of clinical work.

As endodontic postgraduate students on opposite sides of the world, we were both able to learn the key concepts in microscope ergonomics at the beginning of our endodontic careers. This was due to the community created by Dr. Gary Carr, the pioneer of microscopes in endodontics. We were then able to successfully train teams of assistants in private practice, where every procedure is done efficiently and together using a microscope.

This led to lecturing in several countries on how endodontists can use the microscope efficiently. Through video and explanation, attendees were able to see a condensed overview on how we trained and responded to feedback in developing expertise in microscope use. Whilst displaying the end result shows what is possible, the most common question at the end of these lectures asked about a book or other resource where people could learn at home, and reinforce the lessons learned in a 1 h presentation. An online video course (quantumergonomics.com) was created by one author (KY) which led to the idea for this book.

Our hope is that future generations of endodontists will have a simple path in developing efficient microscope skills with their team. By collating all of the thought processes we have learned, this textbook serves as a key reference for learning the lessons that very few people teach.

This is a dedication to our friend and mentor, Dr. Gary Carr, who by teaching microscope skills to endodontists changed our specialty and used those ideas to create a culture of excellence.

The authors would like to thank Drs Ed Kasper and Mike Trudeau for their assistance in reviewing chapters.

East Melbourne, VIC, Australia Kehn E. Yapp
Heemstede, The Netherlands Carlos Aznar Portoles

Contents

1	**Team Culture**	1
	References	7
2	**Operatory Design**	9
	Circle of Influence	10
	Microscope Components and Function	12
	Eyepieces	14
	Binocular Tube	16
	Magnification Changer	17
	Objective Lens	18
	Assistant Binocular Tube and Beam Splitter	19
	Dual Iris	20
	Relationship Between Microscope and Posture	21
	Operator and Assistant Stools and Positioning	26
	Patient Chair and Microscope Positioning	28
	Operatory Design	30
	Back Wall Design	31
	X-Ray Arm Positioning	32
	The Cart	34
	Microsurgery Layout Considerations	36
	Door Locations and Patient/Staff Movements	37
	North/South Orientation and Light/Windows	38
	Reference	39
3	**The Art and Science of Documentation**	41
	Scott K. Bentkover	
	Introduction	41
	Important Photography Basics	42
	Microscope Armamentarium Important for Photography	47
	Getting Started	54
	Ergonomic Photography Skills	56
	Putting It All Together	61
	Video	62
	Conclusion	64
	References	64

4 **Training Principles** 65
Movement Principles 70
References 75

5 **Training Programme** 77
Mirror Change 81
1–2 Instrument Pass 84
Irrigation 91
File Passing 93
Local Anaesthesia 96
Right Hand Anaesthesia 97
Left Hand Anaesthesia 99
Rubber Dam 102
Capsule Applicator 104
Composite Applicator 105
Left Hand Skills 106
Cutting with Scissors/Scalpel 108
Suction Handling and Placement 111
Assistant Stropko Use 113
Taking Radiographs 115
Placing Amalgam 116
Full Procedure Simulation 118
Surgical Training 121
Team Footage Review 122
References 124

6 **Clinical Microscope Endodontics: Consultation and Case Presentation** 125
Before the Appointment 126
Setup and Checklist 126
During the Appointment 128
Patient Management Upon Arrival at the Office 128
First Meeting with the Assistant 128
First Meeting with the Operator 129
Starting Protocol and Patient Examination 130
Intraoral Imaging and Documentation 135
CBCT Study Review 137
Case Presentation 138
After the Appointment 139

7 **Clinical Microscope Endodontics: Non-surgical Treatment** 141
Introduction 141
Before the Appointment 142
During the Appointment 144
Start Treatment 144
Complete Treatment 153
After the Appointment 156

8 Clinical Microscope Endodontics: Surgical Treatment 157
Ivan N. Vyuchnov
Introduction 158
Before the Appointment 161
First Surgical Assistant 162
Second Surgical Assistant 163
Microsurgical Instruments 164
Documentation 165
Positioning 167
During the Appointment 170
Soft Tissue Management 171
Osteotomy 174
Root End Management 175
Canal Preparation 178
Canal Filling 179
Suturing and Wound Healing 181
Suture Types 181
Suture Geometry 184
Clinical Suturing Steps 190
After the Appointment 198
Suture Removal 198
Training 199
References 200

Correction to: Team Microscope Endodontics C1

Closing Remarks 201

Team Culture

1

Abstract

This chapter discusses the assignment of values that are important to establishing a positive team culture. The application to a quality focused, microscope centred practice is presented. Mechanisms of learning and developing expertise are described. Key aspects of a training programme designed for team members to achieve the goals of this culture are covered.

The way you do one thing is the way you do everything—anonymous

Key Points
- Clearly establish a culture of excellence to your team.
- Develop expertise with a clear team training programme.
- Set your team up for success, with the right layout, equipment, and values.

We train as a team to do everything using a microscope for one main purpose: to create a working environment with the highest level of health and safety, efficiency, teamwork, and professionalism, to provide the greatest level of patient care. This ensures that the operator, assistant, and patient are all happy. The goal of using the microscope together for all procedures is possible and realistic for everyone. It is an individual's motivation and commitment to learning new skills that is the limiting factor. The team focus should be the most important, simply because these treatment procedures cannot be done by a single operator. The outcome is that both operator and assistant will become experts in team microscope use. The culture of a microscope-centred practice should focus on excellence, embrace all of the above ideas, and lead to better clinical outcomes.

K. E. Yapp, C. Aznar Portoles, *Team Microscope Endodontics*,
https://doi.org/10.1007/978-3-031-70188-7_1

There is a real and widespread problem for the entire dental profession—physical pain due to the working environment. Musculoskeletal pain in dental professionals has been reported to affect over: 40% of German dental assistants (Haas et al. 2020), 80% of Italian dentists and dental hygienists (Gandolfi et al. 2021), 87% of first-year dental students (Santucci et al. 2021), 96% of Czech dentists (Hodacova et al. 2015), and 93% of Saudi dental professionals (Bakhsh et al. 2021). These observations have been consistently reported over time: pain and physical injury as a result of doing dentistry is very common. These are back, neck, shoulder, and arm injuries. Logic indicates that the reason is clear—a hunched over, forward leaning position to look directly into the operating field (the patient's mouth) over any period of time is bad for your body. Efficiency refers to minimising wasted movement. Being efficient can prevent injury in the workplace. Wasted movements and operational inefficiency are also bad for your body. How is this problem solved? Simple. Don't lean and hunch over, with a bent posture, whilst performing dentistry. Stop putting yourself in harm's way.

Imagine you turn up to the surgery, see patients and perform treatments all day, in a relaxed team environment that does not cause undue physical stress, nor put you at increased risk of injury. Then, imagine you finish your working day, go home and do not have to deal with any physical strain, injury, or pain. All of the lessons in this book will be able to turn imagination into your reality. This has been our reality for several years, in addition to a handful of operators around the world. We created our teams, intensively practised to develop expertise, established the right culture, and were able to embrace these skills, leading to a fulfilling and pain-free work environment.

When you create a culture, whether it is in the workplace or in a personal setting, the two most important components are clearly communicating a clear vision on: (1) what you want your team to be, and (2) what you want your team *not* to be. The first point includes establishing: that all clinical work is done together efficiently using the microscope, that all team members are prepared and proactive for each situation, and that communication is clearly undertaken and understood between everyone in the practice. Everyone's role in the team and expected performance tasks are to be unambiguous; no one should be surprised if their behaviour does not follow their duties within, and for, the team. The second point is equally as important as the first. By being forward about what you don't want, this gives the team more direction on how daily operations should occur. Some of these involve aspects indicating that you do not want your team to be: inconsistent performers, poorly prepared, inefficient in clinical movement and use of time, neglecting patient needs and terrible communicators. Whilst the latter in most cases would not need any explanation, greater clarity is an advantage to everyone. There is no disadvantage with increased clarity of communication, but there are significant risks of putting forward a message with decreased clarity.

By everyone putting the team values first, this set of shared goals will be achieved quicker and easier than a collection of individual efforts. It is very common for anyone with no to little microscope experience, who sees and learns about doing all procedures with the microscope—i.e. exactly what is shown in this book—then quickly conclude to themselves, "this is not possible for me". Such a misconception occurs due to several factors. What actually being internalised is a person who is on day 0 of learning a new skill, visualising another person who is on day 3000 (or greater) of learning that same skill. There is a subconscious comparison of a novice with another person with a highly developed and advanced skillset. There is a further idea of being able to jump from a day 0 learner to a day 3000 learner in a very short space of time, with the internal belief that this quick jump should occur for everyone. Because the day 0 student wants to get to day 3000 in only a matter of days or weeks, this is internally translated to "day 3,000 is not possible for me—ever". You can clearly see the logical disconnect in this way of thinking. Comparing an individual with a highly trained and specialised skillset (team microscope use) with someone without these skills makes no sense. It is identical to comparing a native English speaker who has spent their life learning and practising English—with a person who has never listened to, read, written, or spoken a word of English in their life. It is an invalid comparison.

This message should be emphasised to everyone: incremental gain is how you get better. If you continually focus on improvement and respond to feedback, being an expert is possible for everyone. If you begin with the outcome you want to achieve in mind, setting defined goals will make your journey easier.

The outcomes that both operator and assistant should focus on are health, safety, and happiness for the individual and the team. Shared outcomes that the team should aim for are developing expertise in team microscope use involving the highest level of functionality. This is important because by achieving these outcomes, you can physically and mentally focus on providing the best patient care. By being efficient, you will be able to do the same treatment in less time. You can then allocate more of your day to having time off, or see extra patients in the same amount of time. Mathematically, this means your profitability increases. That alone should be both a personal and professional incentive.

A paraphrased quote by L.D Pankey is: "The best dentists make their patients worse at the slowest possible rate". This underlying message refers to making clinical decisions that do not cause harm to patients, minimise deterioration of their oral health, and also provide long-term benefits.

Here, we take that saying to the next level:

"The best teams of operators make themselves worse at the slowest possible rate".

Clinical dentistry is a perishable skill. The sooner you embrace this idea, the more you will be able to value your clinical career—by enjoying the best moments and realising they, like other good things, must eventually come to an end. Physical injury in dentistry is unfortunately a common occurrence, and it has a lot to do with poor posture and movement. Preserving this skillset not only increases longevity, but also improves the physical and mental execution of the relevant clinical tasks. You are only of value to your patients if you can perform to the highest level in the chairside setting. If you are injured, that does not help patients.

As an exercise, count your likely number of remaining practice years from now until retirement. Ask yourself, how would you feel if you had to stop your career earlier than planned? How many years would be left, if for physical reasons you had to stop practising 5 years earlier? Or even 10 years? If you ask any dentist or dental specialist who had to cut their career short due to workplace-related injury, the common finding would be that their physical condition declined over time. There was no sudden back issue as a result of a one off incident of leaning over in the wrong position. Their health was getting noticeably worse and worse. This means the number of remaining practice years *with a good quality of life* would be decreased even further, prior to an early retirement due to physical injury from working in the dental operatory.

To make the most of, and preserve, a team microscope skillset, we teach the following ergonomic foundation skills.

1. Do everything using the microscope
2. Minimise wasted movements
3. Move the patient and not the microscope

These foundations help achieve the goals of a healthy, safe, and happy environment by ensuring that the team operates efficiently (Fig. 1.1). This means minimising wasted time and movement for both operator and assistant.

The right operatory setup is the only way to achieve these goals. As you will soon learn, the design of your surgery and choice of your equipment is all related to the ability of the team in performing all procedures together, using the microscope, with the highest level of efficiency. If you already have a completed surgery setup, you will find that changing the physical layout and equipment will be far easier than changing bad habits, ways of thinking, established patterns of behaviour, and the culture of your practice. As the operator, you want to set your team up for success. An assistant wants to be part of an environment where everyone is valued, the team thrives on achieving great outcomes, and each member has a fulfilling role. This can be done by ensuring your physical layout is conducive to achieving all of the above goals, and a positive workplace culture is instilled in the entire team.

This book is aimed at changing the practice and behaviour of the reader—whether you are an operator or assistant. Teaching describes what the teacher

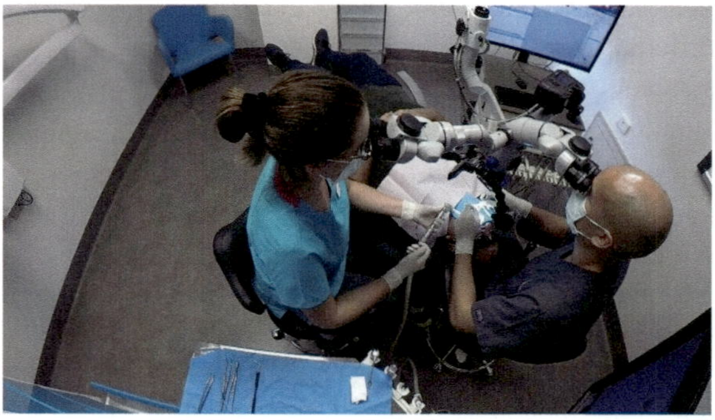

Fig. 1.1 Photo of a real patient scenario where both operator and assistant are working comfortably, efficiently, relaxed, and both are using the microscope to focus on the treatment

does—to follow a teaching plan by using various interventions to meet intended learner outcomes. It does not relate to how the student changes practice or behaviour.

Learning refers to the *relatively permanent* change taking place as a result of extended practice. The aim is to create change—because without change, there is no learning. The ultimate goal of learning is to create expertise, which occurs when training improves skillset development and includes repetition and the immediate opportunity to improve. This repetition is aimed at creating automation. As the reader, the permanent changes following the lessons described in this book are most important. You will only get better if these changes are permanent.

Three stages of learning have been described:

- Declarative knowledge
- Mixed knowledge
- Procedural knowledge

Declarative knowledge relates to facts. The two types are recognition and recall. This involves being able to view equipment, materials, images related to the task at hand and recognise their name, use, and orientation. Recall of objects is the other type, i.e. selecting a Briault probe when asked for one. The strength of declarative memory items increases with use and decreases with lack of use. Without declarative knowledge, learning cannot progress to the next step. This is the foundation for the procedural actions, where mixed knowledge bridges the gap between the two. Mixed knowledge is when declarative knowledge is consolidated into procedural knowledge.

Procedural knowledge is the learning of ordered actions. It is based on declarative knowledge because without it, a logical sequence of steps cannot be followed. An example is: you need to know what endodontic access is and what canals are, before thinking about using files to get patency. You cannot jump straight to patency without finding canals first.

Every procedure is a sum of multiple components. They may be complex, but not difficult. Because these components are sequential, each step relies on the one before it to be correct, for the entire procedure to work. Training compiles declarative knowledge into a proceduralised skill. After reading this book, the objective for both operator and assistant is to develop permanent procedural knowledge so that both can work as a team to use the microscope together—i.e. to achieve the learning outcomes.

Expertise operates using the recognition of stored patterns. As these team microscope skills develop, the amount of information included in these patterns increases. The development of an assistant skillset requires an understanding of the requirements in every situation, and the ability to implement and anticipate them in a logical order. Training stores these patterns in a process called deliberate practice. This "includes activities that have been specially designed to improve the current level of performance". These tasks are aimed to overcome weaknesses and "performance is carefully monitored to provide cues for ways to improve it further". Deliberate practice requires effort and is not inherently enjoyable. This is defined by four criteria (Ericsson and Harwell 2019):

1. The practice involves individualised training of a trainee by a well-qualified teacher.
2. The teacher must be able to communicate the goal to be achieved by the trainee and that the trainee can internally represent this goal during practice.
3. The teacher can describe a practice activity to attain the identified goal for performance and that this activity allows the trainee to get immediate feedback on a given attempt.
4. The trainee is able to make repeated revised attempts that gradually approach the desired goal performance.

What this means, is that it is up to both the teacher and the student to devise and follow a training programme, that can be implemented with the outcome of achieving specific goals with immediate feedback to improve (Fig. 1.2). This is what this book will show you.

One important trait of experts is that they are able to readily adapt to changes in their environment and still achieve the required goals. They do not have a fixed mindset, nor do they have a rigid working protocol that cannot be modified. For any potential adversity, a solution is found to overcome the problem.

This means that the overall training programme will focus on core competency skills because they occur most frequently, in addition to methods to adapt. In endodontics, the treatment procedure generally follows the same sequence that fits a logical progression. Because there is an identical order of movements, the working focus should be centred on the next couple of steps only, not 10 or 15 steps ahead. This is what experts in other domains do (Connors et al. 2011). Knowledge of the learned sequence is important in understanding what is and is not to be expected. Meaning, protocol deviations are only recognised when the protocol is learned. When there is a change in required procedure, alternative steps should be anticipated and encompassed as part of the training process. Examples of alternative

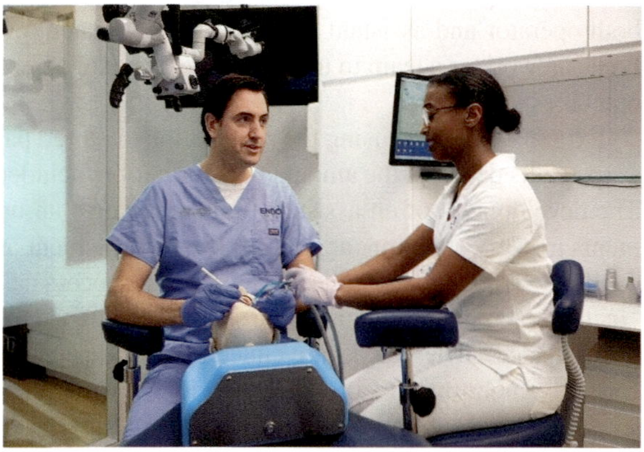

Fig. 1.2 One-on-one training with a new assistant

equipment, materials, or techniques would need consideration in the event that the standard pattern cannot be followed. One example is the matrix band not adapting to the tooth properly. Alternatives in the situation would be using a different matrix (Tofflemire #2 instead of #1), system (sectional or "V-shaped" ring matrix), or the operator holding the matrix by hand. The more readily the team can recognise alternatives and implement them when needed, the greater the expertise of the team.

The bigger picture goal is to create a culture in your practice that is centred on microscope teamwork and efficiency. For each team member, the aim is to develop and implement ways on how to think, how to analyse situations, identify issues, and solve problems. This starts with the basics of individual skills and finishes with a complete understanding of executing the procedural sequence. Once this level of expertise has been established, teaching these ideas, processes, and ways of thinking to the next generation or a new student becomes simpler.

How about the interpersonal goals? The most important part of a working relationship is trust. This is the foundation of every high functioning team; it is an understanding that each member of the team relies upon each other to achieve the team goals. Without trust, there is no cohesiveness, no balance, no shared objective, and no team.

References

Bakhsh HR, Bakhsh HH, Alotaibi SM, et al. Musculoskeletal disorder symptoms in Saudi allied dental professionals: is there an underestimation of related occupational risk factors? Int J Environ Res Public Health. 2021;18(19):10167.

Connors MH, Burns BD, Campitelli G. Expertise in complex decision making: the role of search in chess 70 years after de Groot. Cognit Sci. 2011;35(8):1567–79.

Ericsson KA, Harwell KW. Deliberate practice and proposed limits on the effects of practice on the acquisition of expert performance: why the original definition matters and recommendations for future research. Front Psychol. 2019;10

Gandolfi MG, Zamparini F, Spinelli A, et al. Musculoskeletal disorders among Italian dentists and dental hygienists. Int J Environ Res Public Health. 2021;18(5):2705.

Haas Y, Naser A, Haenel J, et al. Prevalence of self-reported musculoskeletal disorders of the hand and associated conducted therapy approaches among dentists and dental assistants in Germany. PLoS One. 2020;15(11):e0241564.

Hodacova L, Sustova Z, Cermakova E, et al. Self-reported risk factors related to the most frequent musculoskeletal complaints among Czech dentists. Ind Health. 2015;53(1):48–55.

Santucci NM, Jellin J, Davenport TE. Dental and physical therapy faculty collaborate in assessing and educating dental students on musculoskeletal disorders. J Dent Educ. 2021;85(1):53–9.

Operatory Design

2

Abstract

This chapter covers the types of movements and how the operatory design focuses on minimising wasted movements. It describes all of the required functions and features of a microscope to ensure team members can work efficiently and maintain physical health. The correct posture and positioning for both operator and assistant is discussed. A sample operatory plan which encompasses ideal ergonomic design factors is given.

Measure twice, cut once—anonymous

Key Points

- Position all key instruments and materials in the circle of influence.
- Setup microscope components so you can sit upright and work without moving the microscope.
- Limit motions to class 1 and 2 where possible.

Working with the microscope in an ideal environment requires a specialized setup for the clinical team. Several factors influence microscope ergonomics which require careful attention when designing the operatory. In an ideal ergonomic setup, all treatment will be performed from start to finish under the microscope, from consultations to surgical and non-surgical treatment. Even procedures that are routinely not performed under the microscope can be performed very efficiently with the correct operatory design. Clinicians who do not practice this way will often fail to develop ergonomic skills to practice at the highest level.

K. E. Yapp, C. Aznar Portoles, *Team Microscope Endodontics*,
https://doi.org/10.1007/978-3-031-70188-7_2

Circle of Influence

The first principle to consider when designing an ergonomic operatory is to having everything in the "circle of influence". Ergonomics classifies the motions required to perform a task into five categories. A task that can be performed in a single motion will be ergonomically more efficient than one that requires multiple motions. An example of a task that can be performed with a single motion is passing and exchanging instruments like probes, files or mirrors, whilst other tasks like taking a handpiece or bringing the microscope to the operatory field will require multiple motions (Figs. 2.1, 2.2, 2.3, 2.4, and 2.5). The five classes of motions are:

- Class I: only moving the fingers
- Class II: movement from the wrist
- Class III: movement from the elbow
- Class IV: movement from the shoulders
- Class V: movement from the waist

The circle of influence principle states that all instruments and equipment necessary for the operator and assistant are within arm's reach, at *no more than a class III motion* (Fig. 2.6).

This principle will have direct influence on the dimensions of the operatory. In an ideal ergonomic scenario, most endodontic procedures will be performed only with class I and II motions (not taking the hands out of the circle of light during the treatment), and the assistant will be responsible for positioning the instruments in the fingers of the operator, into the field of view. Class III motions will be employed by the operator when reaching to certain instruments, like handpieces, ultrasonics, or Stropko irrigators. Class IV motions will be employed only for bringing the microscope in and out of the field or to change magnification during treatment. Class V motions can be prevented when working in an ergonomically designed operatory.

This reduction on the number of motions will have a direct benefit on stress and fatigue reduction, injury prevention, and will increase competence and enhance teamwork efficiency. The principle of the Circle of Influence places the microscope at the centre of the operating field, and all these ergonomic motions are centred within these circles. It is also designed to allow all the procedures to be performed under the microscope.

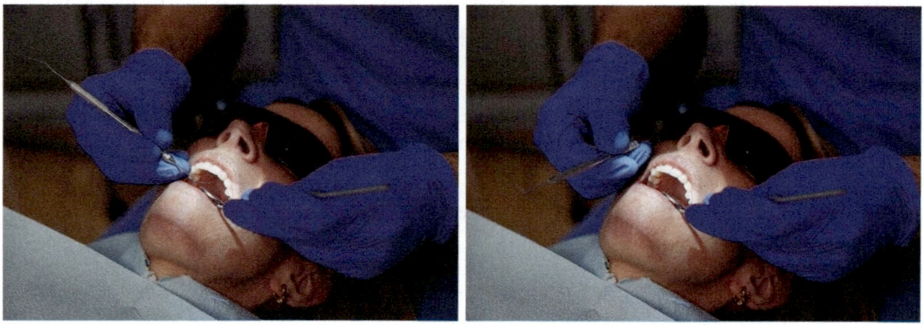

Fig. 2.1 Class I motions—finger movement only

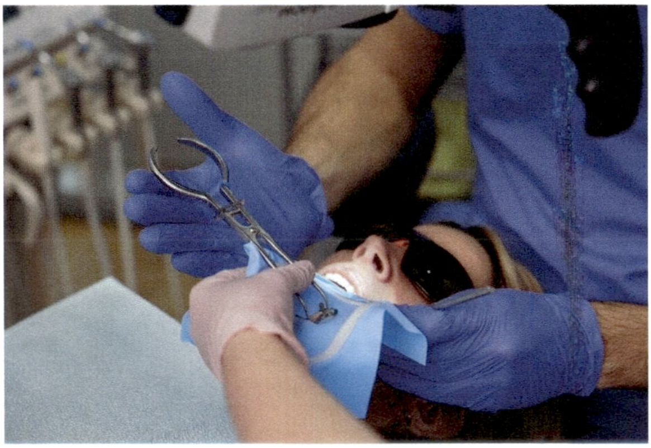

Fig. 2.2 Class II motion—wrist movement

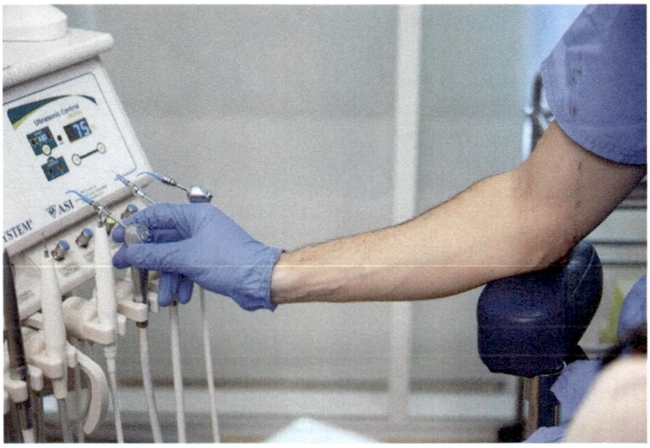

Fig. 2.3 Class III motion—elbow movement

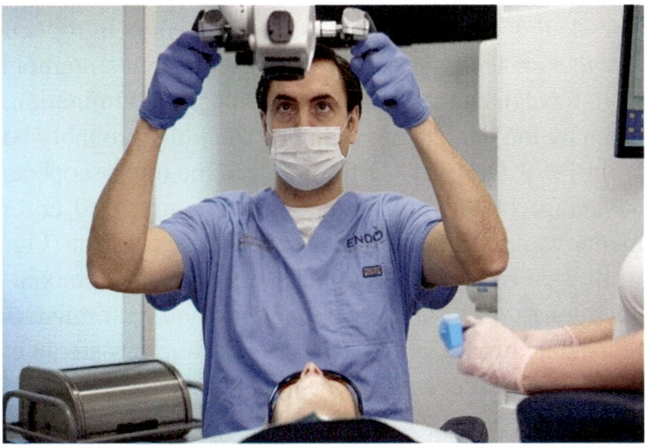

Fig. 2.4 Class IV motion—shoulder movement

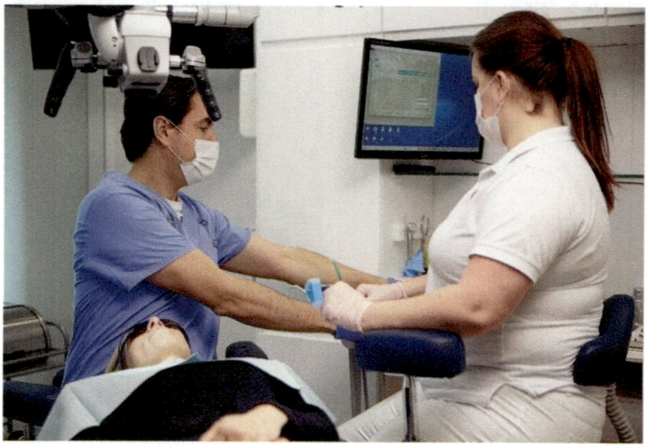

Fig. 2.5 Class V motion—hip movement

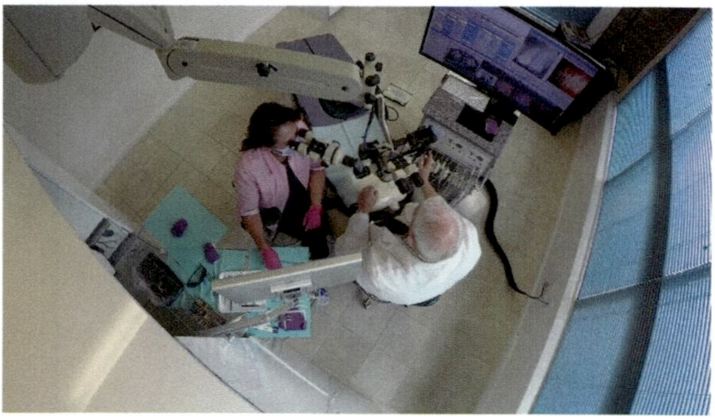

Fig. 2.6 The dental cart and back wall tray are within arm's reach of the operator and assistant, respectively (courtesy Dr. Gary Carr)

Microscope Components and Function

The dental operating microscope was first introduced for dentistry in 1981 by Apotheker. This was a floor model with fixed straight binoculars, one magnification (8x), focal distance of 250 mm and angled illumination. There was very little acceptance by the dental community at the time, probably because of poor ergonomics.

In 1999, Dr. Gary Carr introduced the microscope as we know it today. The main advantage of this microscope was the improved ergonomics. The main improvements were: Galilean optics reducing eye strain (Fig. 2.7), 5-step magnification, wall or ceiling mount, angled binoculars, and coaxial illumination. It allowed the operator to perform most endodontic and restorative procedures sitting down, in a relaxed neutral position. All later microscopes are an evolution of this model.

The microscope components and their specifications have a direct influence on the procedures that can be performed. As already stated, all the procedures should be performed under the microscope. For endodontists, we can easily identify the benefits of using a microscope during an endodontic procedure for the location of canal orifices, identification of caries, or the diagnosis of root fractures. Other procedures like giving anaesthesia (Fig. 2.8), performing restorative procedures or occlusal adjustments are now possible if the lowest magnification of the microscope is approximately 2.2x.

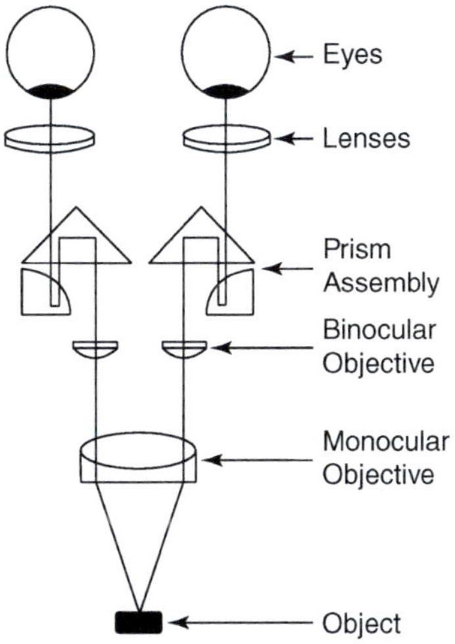

Fig. 2.7 Adapted from Carr and Murgel (2010)

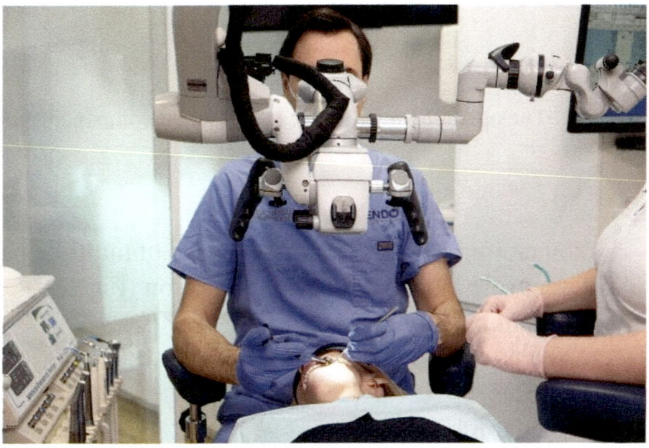

Fig. 2.8 Administering local anaesthesia using the microscope

Working with such a low magnification will allow the operator to have a wider field of view, being able to visualise from the bottom of the nose to the chin of the patient, and stay focused on the microscope throughout the whole procedure. In a set up where the lowest magnification is higher (3x or above), the field of view is so restricted and the operator will have to move the microscope in and out of the operating field multiple times to use the microscope. This will result in a high number of unnecessary interruptions that will increase treatment time, have a direct effect on fatigue and stress and lower productivity. In the long term, it will increase the risk of suffering from musculoskeletal disorders and disability.

Most important microscope specifications:

- Lowest magnification having a wide field of view to see the whole mouth
- Assistant binocular tube
- Binocular extender and 50 mm spacer so operator can sit upright
- Dual iris for increased depth of field

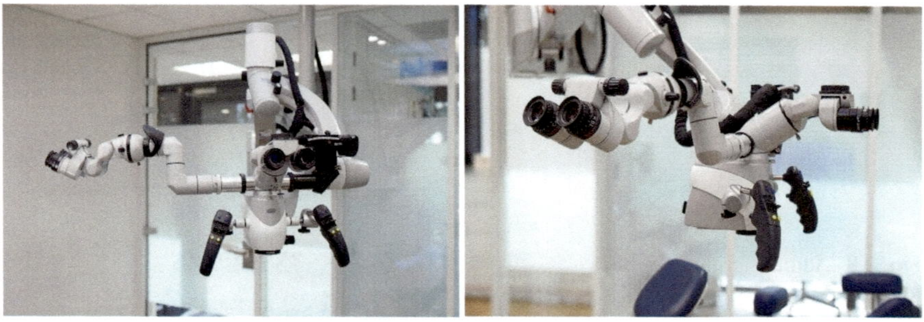

Fig. 2.9 The operating microscope and required components for efficient teamwork

An operating microscope is composed of multiple parts, and each component has a different role in the overall magnification (Fig. 2.9).

The main parts of the microscope are:

- Eyepieces
- Binocular tube
- Magnification factor
- Objective lens
- Assistant binocular tube
- Beam splitter
- Dual iris

Eyepieces

This part is in contact with the operator's eyes. Their function is to magnify the image coming from the binocular tube. Most microscopes are equipped with 10x or 12.5x magnification eyepieces. The larger the magnification factor, the higher the end magnification, and the smaller the field of view. This will complicate performing some procedures under the microscope. For both operator and assistant, eyepieces with a magnification factor of 10x are required to keep the end magnification as low as possible and consequently work with a larger field of view.

Eyepieces are equipped with a ring for diopter corrections, which is a measurement of the optical power of the lens—the degree which it converges or diverges light. The diopters have "0" (neutral), "+" and "-" positions. Each diopter will have to be adjusted for each individual eye, starting at the "0" setting. Special attention has to be given to the dominant eye. A reticle (crosshairs or gun sight) should be used on the camera side eyepiece to help the operator stay centred in the point of interest and help with parfocaling. When a lens is parfocaled, the object is in focus every time the magnification and/or focal length is changed. With correct parfocaling, the operator, assistant, and camera (used for documentation) will be simultaneously in focus.

When configuring eyepieces for the first time, it is critical to have your eyes tested and the correct prescription for corrective lenses applied as needed—if a recent examination has not occurred. This will also provide the interpupillary distance, which needs to be correctly set to ensure both eyes can focus. With your eyes at the right prescription, you will be able to focus on far objects when looking out of the microscope, such as monitors or TV screens, as well as ensure your photos are in focus (see Chap. 3 on Microscope documentation for more details).

Eyepieces are equipped with plastic or rubber eyecups designed to be extended and adjusted to the operator's eyes to shade external light and help visualise the entire field of view. For operators who wear glasses, these rubber eyecups should be retracted because glasses function already as spacers. There is controversy about the benefit of eyecups in clinical microscopy. Whilst there is no doubt about the beneficial effect in fixed microscopy, such as with a laboratory microscope, in clinical microscopy keeping a small distance between the eyepieces and the operator's eyes is advantageous (Fig. 2.10).

This keeps a certain degree of peripheral vision, allowing visualisation of the patient's face, gestures, and looking at distant monitors without having to take the eyes out of the microscope with a class V motion. These small interruptions may seem minor, but when repeated multiple times during the whole course of the treatment and throughout the working day, will result in extra clinical time and fatigue. The recommended setup is to have the eyecups retracted and to have your eyes about 2 cm from the surface of the eyepieces. Vision will not be hindered with this position and you will be able to look out of the microscope by moving the eyes only, not from the hip.

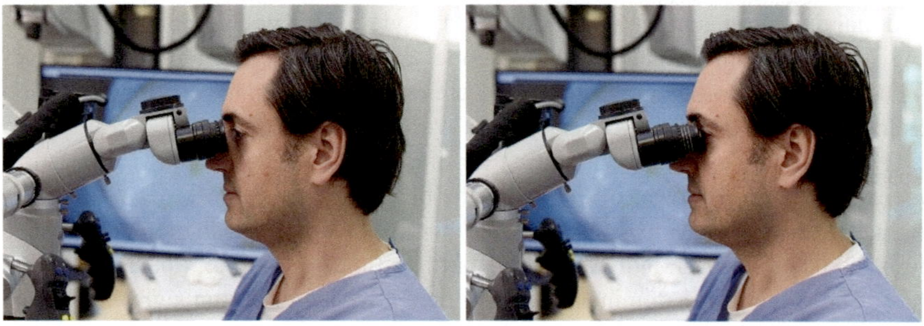

Fig. 2.10 Left: correct eyecup position allowing peripheral vision. Right: incorrect eyecup position preventing peripheral vision, resulting in a hip (class IV) motion to look out of the microscope

Binocular Tube

This is the part attached to the body of the microscope that contains the eyepieces. Inside the binocular tube are a series of prisms that rotate and produce an upright image at the eyepiece. This image is transmitted through two optical paths with slightly different angles, what produces an intermediate image that is magnified by the eyepieces and projected to the exit pupil. Our eye lenses receive this image and project it into the retina. Our brain will then form a three-dimensional image or stereoscopic image. Stereoscopy allows the operator to perceive depth and distances, orientate objects, and work in a precise manner. The binocular tube also allows the operator to adjust the distance between the pupils in order to match the exit pupils of the microscope.

The tube head has a lens with a determined focal distance. The greater the focal distance, the higher the magnification. For example, most Zeiss binocular tubes are equipped with a focal distance of f = 170 mm, whilst other brands, like Global, have a tube with focal distance of f = 160 mm.

The binocular tube has a direct influence on microscope ergonomics. Ideally, the body of the microscope will remain perpendicular to the floor during the whole procedure, independent of the tooth or arch being treated. The binocular tube serves as a "connection" between the body of the microscope and the operator's head. From an ergonomic perspective, the binocular should allow the operator to work in a relaxed posture, with no neck strain. There are three main models of binocular tubes:

- 45° fixed head tube. The operator will have to lean forward and down to see into the eyepieces, with compromise to the ergonomics.
- 0–180° inclinable head tube. This head tube will allow the operator to adjust the viewing angle, remaining in a neutral relaxed position (Fig. 2.11).
- Foldable tube (Zeiss). F = 170/260 mm. This is the most flexible solution when the same microscope is shared by multiple operators. This tube allows different operators or assistants to work at the same working distance but with the head

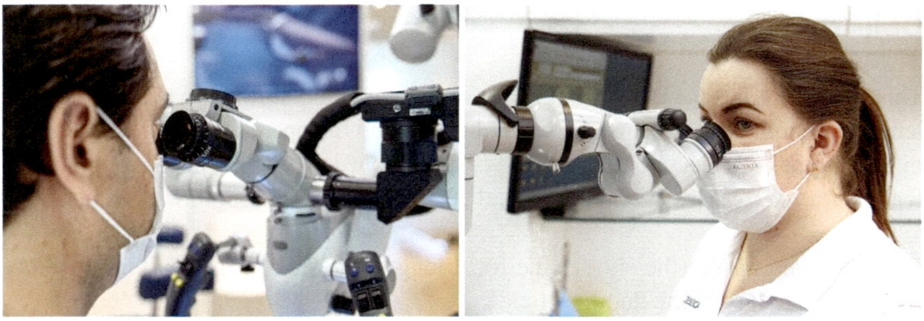

Fig. 2.11 Left: 0–180 degree inclinable tube with a 50 mm spacer. Right: foldable Zeiss tube

tube at different eye levels or to compensate any height discrepancy between operator and assistant. It also offers a 50% higher magnification at the eyepieces with the turn of a knob for cases where higher detail is needed (Fig. 2.11).

Accessories can be added to the binoculars, like spacers and angled extenders, to allow the operator to work at a further distance from the patient or with the microscope at an angle whilst keeping an upright posture. The extender angles the surface of the binocular attachment to face the operator, making it closer to the eyes. Some angled extenders incorporate a beam splitter in the same unit. A 50 mm spacer is required for a 0–180° tube to ensure both the long axis of the microscope and the torso of the operator are perpendicular to the floor, allowing the operator to have a relaxed and straight back, neck, and shoulders.

Magnification Changer

This part is located at the body of the microscope. It allows the operator to change the magnification factor during the procedure. We find typically two types of magnification changers: a step magnification changer or a zoom system.

There are several models of step magnification changers, usually range from 3-step to 6-step. An advantage of having a 6-step magnification changer is that the lowest magnification factor will result in a smaller end magnification and therefore a larger field of view. This will allow the operator to perform all the procedures under the microscope without moving the microscope, thereby increasing team efficiency.

A zoom system is more convenient for the procedure workflow. When turning the magnification knob in a step system, there is a small period of time where the image is blocked. This is not the case when using a zoom system. Any adjustment of magnification will also be much faster with the zoom system. The zoom system can be operated via the handgrip or a footswitch in cases where it is not possible to take the hands out of the working field. The disadvantages of the zoom system are the higher price and the higher potential for mechanical or electronic problems due to the technical complexity of the system.

Objective Lens

This is the first lens of the system and is positioned closest to the patient. The focal length determines the distance between this lens and the operating field. We have two different types of objective lenses: fixed or varioscope lenses. The most common focal lengths for fixed lenses are 200, 250, and 300 mm. The focal length will influence:

- Working distance (the shorter the focal length, the shorter the working distance and the shallower the depth of field)
- Magnification (the shorter the focal length, the larger the magnification)
- Field of view (the shorter the focal length, the smaller the diameter of the working field)
- Depth of field (the shorter the focal length, the shallower the depth of field)

The focal length of the objective lens plays an important role in ergonomics. In an ideal working position, the operator will be able to slide their knees just under the patient chair, keeping the body of the microscope perpendicular to the working plane and maintaining a neutral posture. Too long a focal length might impede the operator to slide under the patient chair, adopting a posterior sitting position, having to lean forward to reach the eyepieces. This will cause muscle strain on the back and neck muscles. Too short a focal length might make it more difficult to perform certain procedures due to the higher magnification and the neck positioned downwards to look into the binoculars (Fig. 2.12).

This can be compensated in some cases by working with a 6-step microscope with the lowest magnification factor (0.33x). Working with a very short focal length will make passing of instruments more challenging due to the limited space available. For most operators, having an objective lens of approximately 300 mm will provide a good balance with an adequate working distance, comfortable height without neck strain, good depth of field, and ensuring sufficient light into the operating field.

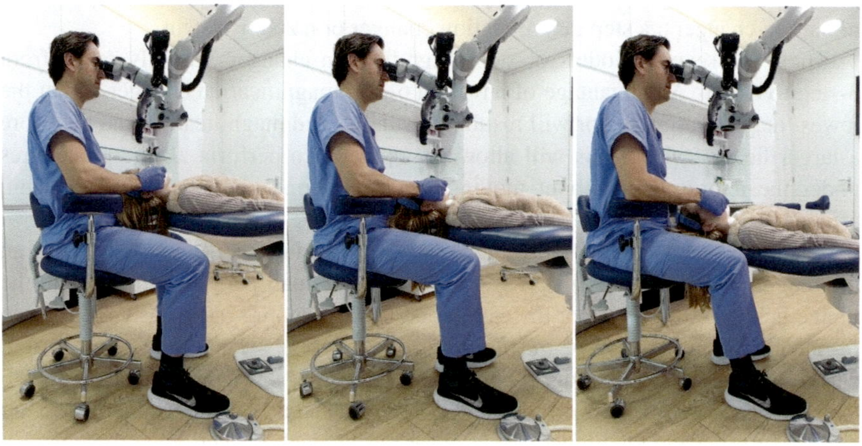

Fig. 2.12 First: too short a focal length. Second: ideal focal length. Third: too long a focal length

To change the working distance during treatment with fixed objective lenses, you will have to unscrew and replace the lens with a different one with a different focal distance. This is unfortunately very impractical and will affect the workflow during treatment. This is one of the reasons why a system where the working distance can be adjusted (varioscope) can be very useful, especially for situations where the focal plane changes during the procedure. In operatories where the microscope is used by different operators, these varioscope lenses can be adjusted at different focal distances to correct any discrepancies. The varioscope lenses have a zoom lens, allowing you to change the focal length that is adjusted manually by turning a knob or electronically by pressing a button in motorised microscopes. The same magnification, working distance, and field of view principles apply to this type of lens systems.

As already discussed, if the microscope allows for a lowest magnification of around 2.2x, we will be able to perform all the procedures under the microscope. Understanding how Total End Magnification is calculated will help the operator to make the right choice when choosing a microscope and its components. The Total End Magnification is calculated with the following formula:

$$\text{Total End Magnification} = \frac{\text{Binocular Focal Length}}{\text{Objective Lens Focal Length}} \times \text{Magnification Factor} \times \text{Eyepiece Magnification}$$

Calculating the overall magnification is multifactorial where some of these variables can be readily changed during a procedure. More importantly, when wanting to have the lowest magnification and the widest field of view, if any one component has too high a value, compensating with lowering another value will be limited and may not provide a low enough magnification, for example, for a Zeiss microscope with a 250 mm objective lens, 170 mm binoculars, 10x eyepieces, and lowest magnification factor of 0.4x, total magnification is 2.72x which is too high to visualise the entire mouth. The only way to correct this is to change to a 300 mm objective lens.

Assistant Binocular Tube and Beam Splitter

The assistant binocular tube is needed for achieving the minimum level of microscope teamwork, allowing the assistant to see exactly what the operator sees. This can only be an advantage for the team where both members have clear and unrestricted vision. It is attached to a beam splitter, positioned at the upper body of the microscope, which splits the beam into two pathways for each eye. On one side, it reflects the image to the assistant binoculars and the other side to the camera. The assistant will be able to sit muscularly at rest during all procedures and perform all tasks more precisely (Fig. 2.13). Some examples of the tasks where the assistant can take advantage of working under the microscope are:

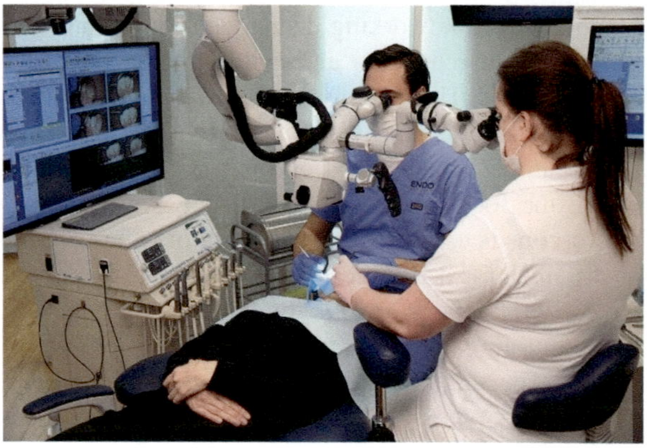

Fig. 2.13 Assistant engaged in the procedure by looking in the microscope through the assistant binoculars

- Precise suction during surgical and non-surgical procedures
- Use of Stropko irrigator during access opening, canal troughing, or restorative procedures
- Documentation under the microscope
- Anticipating procedural steps

The configuration of the assistant binocular tube is critical. A dual axis tube with inclinable binoculars will allow for precise placement to suit the assistant's posture and to perform all the procedures comfortably, as opposed to situations where the scope has a single axis and/or the binoculars are fixed. Many operators may not understand the benefit of working with an assistant with a separate binocular tube. The main benefits are:

- Enhanced teamwork and practice focus
- Increased clinical competence
- Increased efficiency
- Retention of top assisting talent
- Less guesswork on the part of the assistant
- Preservation of the assistant's physical health
- More accuracy in the performance of critical assisting functions
- Promotion of a practice culture centred around clinical excellence
- Job fulfilment

Dual Iris

Control of the aperture opening affects the depth of field and amount of light to the binoculars. Depth of field is the distance between the closest part of the image and the furthest part of the image that are in acceptable focus. Closing the aperture increases the depth of field and is an advantage for clinical procedures. Conversely, opening the aperture decreases the depth of field and limits the amount of the image in focus (Fig. 2.14).

Fig. 2.14 Left: open dual iris allowing more light to pass, decreasing the depth of field. Right: closed dual iris reducing the amount of available light but advantageously increasing the depth of field

Microscopes are not typically manufactured or sold with the ability to change the aperture to the binoculars, meaning the default option has the largest opening, most amount of light and narrowest depth of field. The dual iris is a piece of equipment that overcomes this limitation by giving the operator the ability to change the aperture to both binocular paths (hence the term "dual"). In photography terms, changing the aperture is termed changing the "f-stop" or "f number". By closing the dual iris, the aperture decreases light to the binoculars, increases depth of field, and slightly increases noise of the image. A greater depth of field is also an advantage for photography—see Chap. 3 for more information. Although many may consider this to be optional equipment, a dual iris is essential for working efficiently—by not having to move the microscope when the image is out of focus, in addition to being able to take world class photos and documentation.

Relationship Between Microscope and Posture

Introducing the microscope in the dental office requires a lot of careful thought. Most problems when adapting to the microscope are often related to an incorrect setup, incorrect ergonomic skills and an incorrect posture.

Working with the operating microscope allows the operator and assistant to work in a relaxed upright position, without having to twist, bend, or lean forward during the procedure (Fig. 2.15). The torso should be straight and perpendicular to the floor, which is parallel to the long axis of the microscope. Avoiding muscular tension and stress will reduce the risk of suffering musculoskeletal disorders and disability.

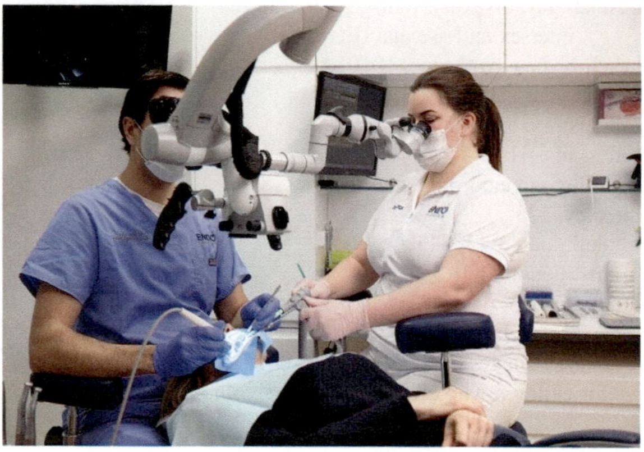

Fig. 2.15 Operator and assistant working in a relaxed upright position

Every time you examine or treat a patient in the chair, a starting protocol should be repeated, ensuring you and your assistant are properly setup and positioned. In chronologic order, the steps for the preparation to work with the operating microscope are:

1. Operator positioning
2. Patient positioning
3. Microscope positioning
4. Interpupillary distance adjustment
5. Parfocal adjustment
6. Fine focus adjustment
7. Assistant binocular adjustment

Ergonomics focuses on the motions applied to achieve any task. Any task should be performed with the least number of motions in order to reduce physical load. Endodontic procedures are often very time consuming and developing a correct working posture will reduce fatigue and lower the risk of suffering musculoskeletal injuries. The operator should try to acquire a working position where muscle tension is kept to a minimum.

It is generally accepted that the correct position is where the hips of the operator are 90° to the floor, the knees are 90° to the hips and the forearms are 90° to the upper arms. This helps keep the muscles at rest. Armrests help to sit straight whilst providing extra support to keep the weight from the neck, shoulders, and upper body, preventing spasms of the trapezius and sternocleidomastoid muscles. Forearms should rest comfortably on the armrests of the operator's chair. Both operator and assistant should have stools with armrests. The assistant is required to have left- and right-handed skills, so sitting on a chair with correctly positioned armrests will be crucial.

A correct posture is that one where both the operator and assistant sit with their back in a neutral position, erect and perpendicular to the floor. The neutral spine position is where your neck and back are under the least stress and strain, with the

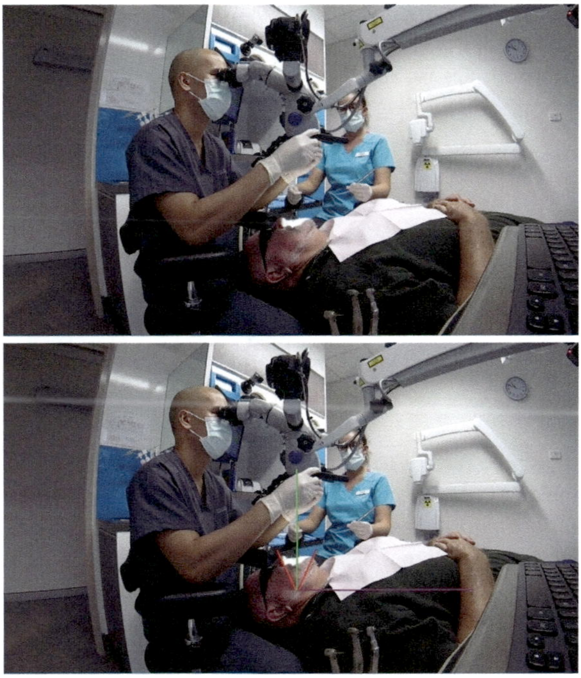

Fig. 2.16 Ideal routine patient position, lying flat and parallel to the floor. Green line showing microscope long axis perpendicular to the floor and patient lying flat (purple line). The microscope long axis bisects the angle between the maxillary and mandibular occlusal planes (red lines), giving the maximum view and most amount of light to the mirror

weight on your ischial tuberosity (or "sitting bones"). The natural lordosis of the back will be supported by the lumbar support of the chair. The armrests will be set to a position supporting the shoulders—where it is not too high—and the hands will help determine the level of the patient's mouth. Eyepieces will be slightly inclined to an angle that the head and neck can sustain comfortably, and the eyes are looking slightly below horizontal.

When reclining the patient for the first time, the most important and efficient part of your starting protocol is to preset the patient chair into the same position. The parameters to set are the angle (flat and parallel to the floor) and the height (patient's mouth in line with the hands when the forearms are resting on the armrests and parallel to the floor) (Fig. 2.16). This reclined position will be used for a majority of patients, regardless of the tooth or the area of the mouth being treated. By setting a consistent position, it saves time getting into position (the preset is selected with one button), gives you a reproducible environment to develop familiarity, helps train muscle memory, inducing automation and develop expertise. The patient should lie comfortably on the chair, with the head, neck, and shoulders supported by a visco-elastic ("memory foam") pillow and chair topper with the head close to the operator's lap.

The ideal patient position is to ensure their head is slightly hanging off the top of the headrest by about 5 cm. This supports the patient's neck and ensures the operator is sitting close enough to the patient, preventing any leaning forward to reach the patient's mouth. For shorter patients, or those who slouch in the chair, you will need

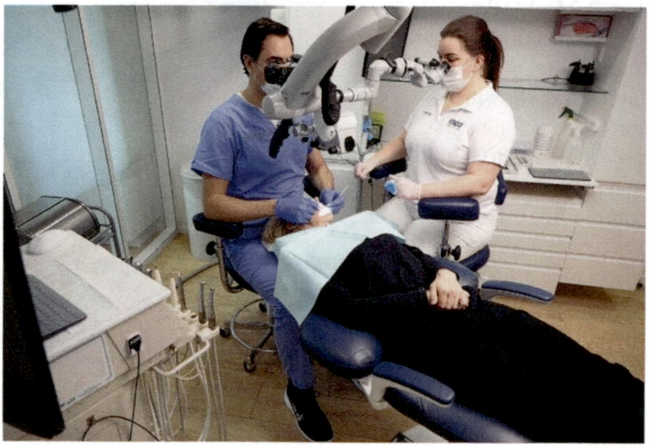

Fig. 2.17 Operator and assistant with arms supported, hips and knees at 90°

to ask the patient to slide up to the top of the headrest. A very common mistake for inexperienced operators is to keep the patient in the wrong position—significantly far down in the chair and far away from the headrest. This error is compounded by adapting to the patient's incorrect position, rounding the shoulders and bending at the hip. Just like all other compliance issues, if the patient adjusts to the wrong position, move the patient back to the right position.

The operator's knees should be at 90° and be positioned either side of the patient's head. This allows the knees to effortlessly rotate the patient chair. For a tall operator with a longer torso and upper leg, the patient chair may be slightly raised with the knees slightly further forward, in line with the patient's shoulders. This still allows the operator to sit straight without leaning forward at the hip and rounding the shoulders (Fig. 2.17).

With the patient reclined and operator's forearms resting on the armrests and parallel to the floor, the patient chair height should be checked before bringing the microscope in position. The mouth should be at the level of the hands, where all procedural movements can occur without significantly changing the height of the forearms.

When the operator has the patient in the right position, the microscope will be moved into the operating field, with the long axis of the microscope body placed perpendicular to the floor. *The first rule to remember in order to work effectively with the microscope is that any small positioning adjustments will be performed by moving the patient, not the microscope.* Moving the microscope continuously will break the natural flow of the procedure and these interruptions will add wasted time, fatigue, and stress. During any procedure, the microscope will be moved only to be brought in and out of the operating field and the only adjustments made would be focusing and magnification change. A chair that can freely rotate will be used to accommodate the patient without having to move the microscope. The lateral movements are performed by moving the chair with the operator's knees. The chair is moved in an east-west direction and the patient's head is rotated in the opposite direction to keep the area to be treated within the field of view.

An important aspect to take into account when positioning the patient is the working distance of the microscope objective lens. Modern microscopes incorporate varioscope objective lenses that allow the operator to easily adjust the working distance to the ideal height in any patient and in area of the mouth to be treated keeping the microscope always at the same position. In cases where the microscope is equipped with lenses with a fixed focal distance, having different lenses with different focal distances will allow the operator to exchange them accordingly in every specific situation. Having to replace lenses during a procedure adds clinical time and interruption. A variofocus system is therefore a much more efficient system in the clinical practice.

Once the rough positioning of the patient is completed, the interpupillary distance will be adjusted. The binoculars have two separate eyepieces, each of them with its own circle of light for each eye. The interpupillary distance can be measured by an optometrist. This value is then set at the binoculars. If this value is unknown, as a temporary measure, both eyepieces will be moved apart from each other and then slowly moved back together until only one circle is observed with both eyes.

As already explained, the patient is positioned completely flat and parallel to the floor. This is the most efficient position for non-surgical endodontics where most procedures are performed with indirect vision using a mirror placed at 45 degrees to the tooth to be treated. With the mirror at this angulation, you will have the maximum amount of light directed to the tooth and canals, and the greatest amount of surface area visualised on the mirror. During non-surgical treatment, this position is constant all the time. When the patient is lying flat, the long axis of the microscope being perpendicular to the floor should be the bisecting angle of the maxillary and mandibular occlusal planes (figure). This is the most consistent and reproducible position giving the greatest view through the microscope. On the contrary, during endodontic microsurgery, the position of the patient will be adjusted depending to the tooth and part of the surgical procedure to be performed.

The next step of preparation is parfocal adjustment. This is a crucial step to reduce interruptions during treatment. When working parfocal, the image will stay in sharp focus when the magnification is changed, with no need to refocus at every magnification step. Working parfocal is not only important for the operator but also for the assistant binoculars and for any camera installed. A camera will be parfocaled when the camera focus and the eyepiece are in focus simultaneously. The fine focus should be set at an intermediate position (at this position it will allow a wider range of focus). The details regarding parfocal adjustment are explained in Chap. 3.

The last part of preparation for working with the microscope is the adjustment of the assistant binocular tube (Fig. 2.18). Once the operator is in position, both hands are placed on the microscope, holding it in position. This is a signal to the assistant that it is ready for any adjustment to the binocular tube as required, regarding the horizontal axis of the tube, or the vertical axis of the binoculars.

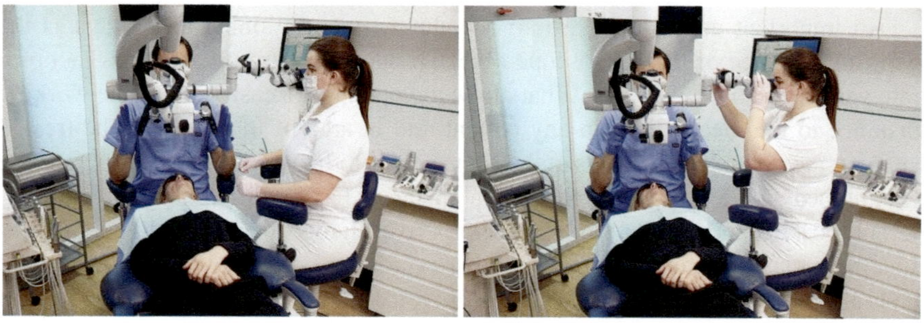

Fig. 2.18 Adjustment of assistant's binocular tube

Starting Protocol Steps

- Adjust the stool so the knees, hips, and elbows are at 90°.
- Preset the patient chair to be flat and parallel to the floor, with the height of the patient's mouth in line with the forearms.
- Recline the patient so their head is slightly hanging off the top of the headrest.
- Bring the microscope into the operating field, with the long axis directed perpendicular to the floor.
- Parfocal, check the interpupillary distance, set fine focus.
- Adjust the assistant binocular tube as required.

Operator and Assistant Stools and Positioning

A correct posture during work is fundamental to avoid musculoskeletal injuries. Endodontic treatment is often a long procedure where a lot of repetitive motions are employed. The longer an operator and assistant practise, the higher the risk of suffering injuries, especially when basic ergonomic principles are not taken into account.

There are multiple models of operator chairs in the market. Whilst small differences among them might be determinant for the final choice (material, colour, price), there are a few crucial principles that must be taken into account.

The chair has to be comfortable; all the components should be adjustable, such as the height, position of lumbar support, and armrests. The design has to allow the operator and the assistant to sit relaxed, in a neutral position. It should have two movable armrests, that when used correctly will eliminate shoulder and neck pain caused by spasms of trapezius and sternocleidomastoid muscles. Armrests must be able to be adjustable both in width and height.

The armrests will be placed in a position where the operator and assistant can rest their forearms at a 90-degree angle with the upper arms. The operating field

Fig. 2.19 The half-circle assistant stool should be avoided because it results in a forward leaning position, leading to back strain

(patient's mouth) should be at the same plane as the operator's forearms. The assistant will usually sit slightly higher than the operator, resting both arms comfortably on the armrests. Half-circle design assistant stools should be avoided because they do not allow the assistant to sit straight, forcing a forward leaning position (class V motion) (Fig. 2.19).

The operator position for all endodontic procedures, surgical and non-surgical, will be at a 12 o'clock patient position. If you consider an overhead view, the patient's head is at 12 o'clock and their feet are at 6 o'clock. With a correct setup, a swivelling patient chair and a properly positioned microscope, there will be no need to change from this position. Any change in the position during the treatment will cause interruptions, waste clinical time, and create an inefficient work environment.

The position of the assistant will be determined by the circle of influence. The ideal position is at approximately 3 o'clock for a right-handed operator. From this position, the assistant will be able to have a wide circle of influence and perform all tasks correctly whilst still keeping a relaxed upright position, without the need to bend or twist the body.

One of the main tasks of the assistant is suctioning during treatment. Ideally, the suction is held with the left hand with the elbow placed on the armrest which facilitates accurate positioning. It is important to avoid severe pronation of the wrist whilst suctioning because it may cause stress, pain, and in the long term, potential injury. There are 90 degree suction tip holders designed to maintain a neutral wrist position.

During surgical procedures, the position of both operator and assistant will remain the same. As discussed in the surgical chapter, it is essential to work with a second assistant. This assistant sits opposite to the first assistant, facing the operator and patient at a 9 o'clock position. The posture is the same as operator and first assistant, sitting up straight, muscularly at rest, resting both arms on the armrests. The circle of influence is slightly larger, around the instrument tray and handpieces, being able to reach the operating field without having to twist the back or bend. This allows the assistant to exchange the instruments and materials from the operator just using class I, II, and III motions.

Patient Chair and Microscope Positioning

The patient chair and microscope position play a fundamental role in working efficiently with the microscope. Ideally, the patient chair has to be designed for patients to withstand long procedures, where it is equipped with extra padding and support so they can lie down comfortably. A few technical requirements have to be taken into account. The chair operating controls have to be accessible for both the operator and the assistant, and they must be designed to be used preferably without having to interrupt the procedure workflow. The main advantage of using a foot pedal is that it allows the operator and the assistant to adjust the chair vertically without having to take the hands out of the circle of light, whilst the main disadvantage is that adds up an extra component to the setup. A control panel on the side of the chair is a good option when no foot control is possible or desired.

The microscope should stay in a fixed position throughout the procedure and any adjustments should be done just by moving the patient chair or asking the patient to slide up or down in the chair. Ideally, a chair should rotate at least 60 degrees in the east and west directions. In addition, the vertical movement should be perpendicular to the floor, not on an angled vertical axis, as this would move the operating field out of the field of view if the patient is moved up or down.

The ideal patient headrest is narrower than the patient's head, allowing the operator's knees to be placed either side of the head. Some patient chair headrests may not allow the operator to sit in a neutral position because they are too wide, making the operator sit further back from the patient. Holding this class V position for extended periods of time will result in increased muscle fatigue. The patient's head should rest on a soft pillow (memory foam) that helps with maintaining a comfortable position during the treatment, facilitating any required head movement whilst keeping good support.

Another chair design aspect to be taken into account is the chair weight allowance and dimensions. The chair has to be able to withstand the weight of patients with a large build. The upper part of the chair has to be also wide enough to give support to the patient's shoulders whilst resting, but at the same time needs to take into account the position of the assistant. A chair that is too wide will force the

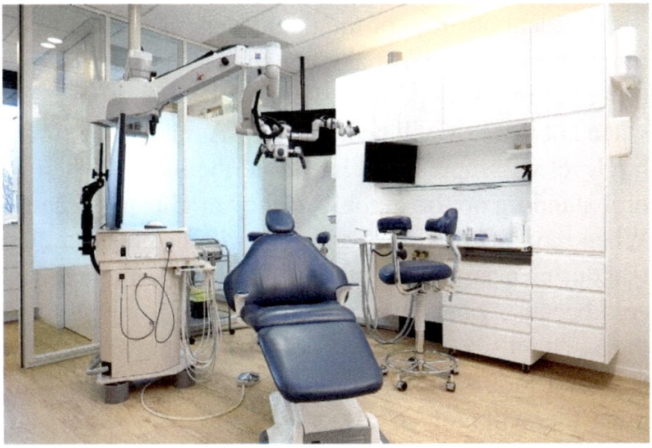

Fig. 2.20 View of an operatory with a ceiling mount microscope

assistant to work further from the microscope and potentially compromise the neutral sitting position.

The microscope position will have a direct influence on the ergonomics of the operatory. When first introduced in the 1990s, many operatories were not designed to adapt to microscope endodontics. Spaces that were too narrow or cabinets placed in the wrong position did not allow for correct placement of the microscope, which negatively affected the ergonomics of the operator, assistant, and the patient. This is one of the reasons that may have contributed to the slow adoption of the microscope in the practice.

The three main microscope mounts are floor, wall, or ceiling. Floor mounts are indicated in situations where the microscope has to be moved between different rooms. The wheels of the floor mount allow for a convenient transport within operatories. The main inconvenience of a microscope with a floor mount is that it occupies a physical space in the operatory, what may compromise the remaining space for the rest of the components of the operatory (chair, cart, operator, and assistant stools). It also restricts the east-west movement of the chair during treatment and restricts a patient's access into the patient chair. Floor-mounted microscopes are not recommended and we advise you to avoid this type.

Ceiling and wall-mount microscopes do not occupy any floor space of the operatory and can be brought easily in and out of the working field, ideally only using a class IV motion (Fig. 2.20). One of the main aspects to consider when deciding the position of the microscope is the distance to the operating field, in both vertical and horizontal planes. If the microscope is placed far from the operating field, the arm will have to be placed in a fully extended position, reducing the range of motion of the microscope. The microscope arm has to allow for some degree of motion in order to allow the operator and assistant to adapt to any situation or patient and help keep a neutral sitting position. When taking working length (or similar) radiographs, the arm should be able to move in a completely horizontal movement to allow the X-ray arm to be brought into place. If a vertical movement has to occur to move the microscope arm, this will disrupt the focus, meaning that once the microscope is moved back into position, parfocaling will have to be unnecessarily repeated.

The ideal position of the microscope is above or behind the operator, either wall or ceiling mounted. As already stated, in this position the operator will be able to bring the microscope in and out of the operating field with a class IV motion and when not in use it will be kept away. A microscope that is mounted over the patient's hip level will represent a hazard to the patient when getting in and out of the chair or if standing up too quickly as they might strike their head against the microscope. When mounted in this position, it will also restrict the motions of the second assistant during a surgical treatment. The height of the mount should be carefully considered because when standing up, the microscope arm will be positioned right above the assistant's head. Some microscope arms are designed such that a rear ceiling or rear wall mount will be in the way of the assistant binocular tube. In some situations, this will prevent proper team microscope use and a ceiling mount above, or to the side of, the patient will be required.

Operatory Design

Careful planning of the surgery dimensions and layout are such a critical component to proper microscope use. Of all major physical factors, poor operatory design is the limiting factor to team efficiency because it dictates how your team can behave and what is or is not possible. The location of the operatory in relation to the whole office, determines where and how staff can circulate and the access points for patients. Details like the operatory width or the door locations can have a huge impact on how simple procedures are performed. A minimum operatory width should be 14 feet (approximately 4.3 m), with 5 feet (approximately 1.5 m) on the operator's right side for the second surgical assistant. This also means to allow for this amount of room, the patient chair should be offset from the centre and not be placed in the middle of the operatory. Extra space on the first chairside assistant side is not of any added value, meaning an asymmetrical operatory is required for the highest level of efficiency.

A main consideration is to separate the entire office into two types of areas: the patient zone and the staff zone. This reduces overlap of circulation space and minimises everyone's distance needed to travel. Another factor is to avoid areas of wasted space, such as corridors. If the office footprint is large, or awkwardly shaped, this may be unavoidable. If you count the minutes in a day that the corridor is used, namely patients entering and leaving the operatory, the time utilised is very minimal. Reducing the distance (and time) required for staff to move throughout the office, for example, from the operatory to the sterilisation room, will minimise waste and aid in efficiency. Quality of space being used is just as important as quantity of space. In a microscope-centred operatory, corners of the room have little to no value. They are bad for storing items because of the distance and time required to travel and retrieve goods. Equipment cannot be mounted in the corner of the operatory because they will be non-functional. The greatest value in a corner is placing operatory doors.

Video link on operatory design:

Back Wall Design

One of the most important factors influencing the ergonomics of an operatory is the back wall design. This design needs to take into account the circle of influence of the assistant and all the elements that form part of it: assistant tray, suctions, mouse and keyboard, or monitor (Fig. 2.21).

The assistant tray has to be large enough to allow placement of all instruments required to complete the procedure. It should be able to slide in and out of the wall to accommodate to the needs of the assistant during treatment. The assistant has to be able to maintain an upright sitting position, muscularly at rest, and be able to reach all the instruments with no more than a class III motion whilst still at the microscope, suctioning or typing notes during the procedure. An important aspect to take into account is the shelf height where the tray and other instruments are placed. This height should be determined by the position of the armrest of the assistant's stool.

The monitor has to be placed at eye level in a direct viewing angle, in such a way that it can be visualised even when the assistant is working at the microscope using the assistant binocular tube, without having to move or turn the head. Placing the monitor out of this line of vision will force the assistant to look out of the microscope, break the workflow and causing interruptions. A monitor holder with an extendable arm facilitates monitor adjustment at the desired angle and allows the assistant to put it away when not in use.

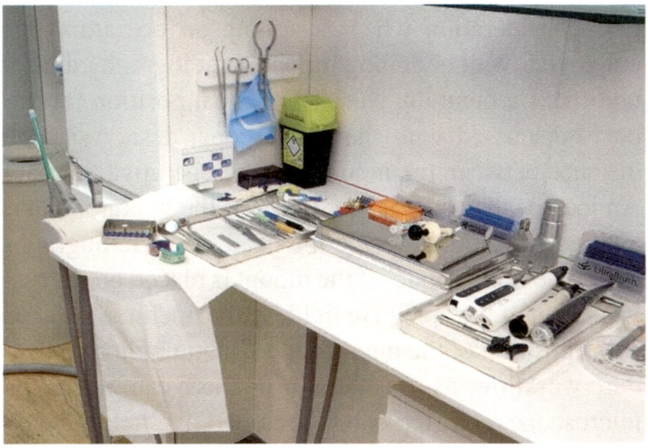

Fig. 2.21 Back wall design with adequate room for suctions, Stropko irrigator and sliding assistant tray

The suction holder should also be incorporated into the back wall, ideally behind the operator at the corner of the assistant tray. From this position the assistant will be able to reach it without having to bend or twist and easily bring it in and out of the operative field. A suction holder should include two suction units (saliva ejector and dedicated endodontic suction tips) that can be used for both surgical and non-surgical treatments. In some cases, a third suction can be added although in certain designs there might be little space for it.

An addition to the suction holder is a Stropko irrigator. This attachment is mounted on an air/water syringe that allows the assistant to have water or air readily available that can be used in specific parts of the treatment such as:

- External bur irrigation during access
- Mirror clearing from water and debris during access
- Debris removal from pulp chamber when troughing or removing instruments
- Crown margin preparation indirect irrigation
- Other restorative procedures

Although an air/water Stropko irrigator seems to be the most efficient option, often water remains inside the Luer-lock tip that make air-only (dry) procedures impossible. For this reason, and whenever possible, a second air-only Stropko irrigator should be added to the assistant platform (next to the air/water one).

The Stropko irrigator can be found in multiple lengths. For the assistant side, the longest version (3.5 in.) should be employed because it will facilitate access to the operative field, particularly on the opposite side of the mouth, whilst maintaining a correct and neutral wrist/elbow/arm position.

Other additions to the assistant tray can be an amalgamator, a composite heater, cordless rotary handpieces or warm gutta percha systems. The same principle applies for all of them: they should be included within the circle of influence of the assistant.

X-Ray Arm Positioning

One consideration when designing an operatory is the radiographic equipment employed. Wall or ceiling mounted radiography units are the most common in dental offices around the world. The ideal position for the X-ray arm mount, as previously discussed with the microscope, deserves careful thought. First of all, the distance between the mount and the operative field needs to be taken into account. An ideal wall mount is that one that the operator can easily reach, ideally just with a class III motion to bring the head of the radiographic device it in and out of the field. The same applies if the mount is placed on the assistant side where it is brought in and out of the operative field for the operator. A second aspect to take into account is the position of the microscope. Taking radiographs will always disrupt the procedure workflow, as the operator and assistant need to stop the treatment, move the microscope, and step out of the operative field. Ideally, the microscope should not

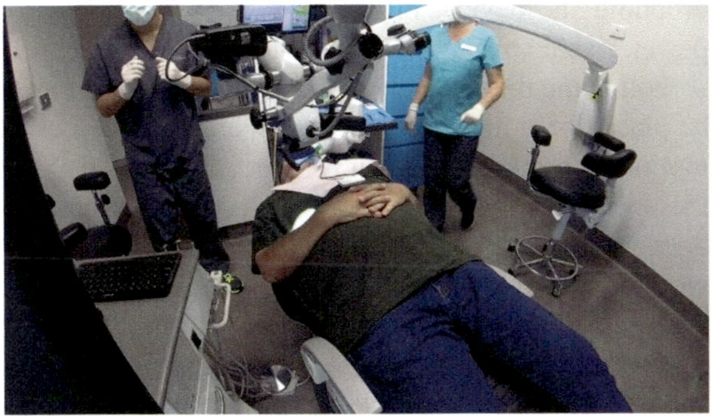

Fig. 2.22 X-ray arm mounted on the assistant wall

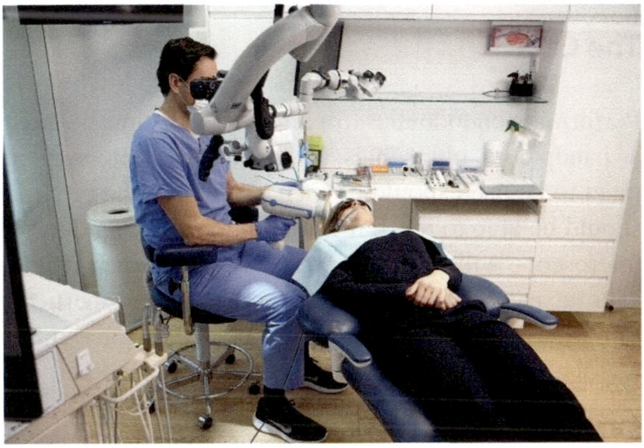

Fig. 2.23 Use of a handheld radiographic device

have to be moved from the working position whilst taking a radiograph; otherwise, re-adaptation to the original position will be more inefficient. This means the microscope starting protocol will have to be repeated multiple times during a procedure. In order to avoid this problem, the arm of the microscope and the radiography arm should follow different paths to the operative field. The microscope should be moved horizontally and not vertically to allow for the X-ray arm to take the radiograph. If any vertical movement occurs, then parfocaling and readjustment will need to occur after the radiograph has been taken, wasting more time and movement for the team. A final aspect to take into consideration is the entry and exit route for the operator and assistant. The arm of the radiograph should not block this route, in order to speed up the process. In most cases, the ideal position for a wall-mounted X-ray arm is on the assistant side wall (Fig. 2.22).

An alternative to the wall mount are the handheld radiographic devices. With this type of device, the operator (or the assistant) is the one holding the device whilst taking the radiograph (Fig. 2.23). A lead-impregnated acrylic shield protects the operator from direct and scatter radiation whilst an internal proprietary housing encases the X-ray tube to block radiation leakage. The main advantage for this type of device is that no installation is required, and there is no need to leave the operative area whilst taking the radiograph. The main disadvantages are: risk of the dropping the device, higher cost when compared to wall-mounted and battery power, risking unexpected power loss.

Following the principle of the circle of influence, when not in use, the device can be placed on a holder next to the operator, which can be reached using a class III motion. Alternatively, it can be kept at the assistant's platform who will hand it to the operator (or take the radiograph). During image capturing, if the microscope does not impede the path of the X-ray arm, the microscope remains at the same working position and the operator or the assistant swivels the patient's chair in an east-west direction in order to be able to face directly the region to be examined. The patient is asked to turn the head to facilitate image acquisition. From an ergonomic perspective, handheld devices are preferred to wall mounted.

The Cart

Performing endodontics to a high level requires the use of multiple instruments such as handpieces, ultrasonics, and Stropko irrigators. As already discussed, any equipment to be used should be included within the circle of influence and the operator should be able to reach it using no more than a class III motion. Working with the microscope will affect the position of these instruments, making some systems like over the patient or rear delivery solutions very inefficient. The best ergonomic solution is the use of an operator cart, placed on the right side (of a right-handed operator) where all these instruments can be placed and controlled with one foot pedal (Fig. 2.24).

Some considerations need to be taken when deciding the type of cart or the instruments to be included. The first consideration is the height of the cart. It is important that the instruments are positioned at the same height as the forearm of the operator. If the instruments are at a higher or lower plane, it will force the

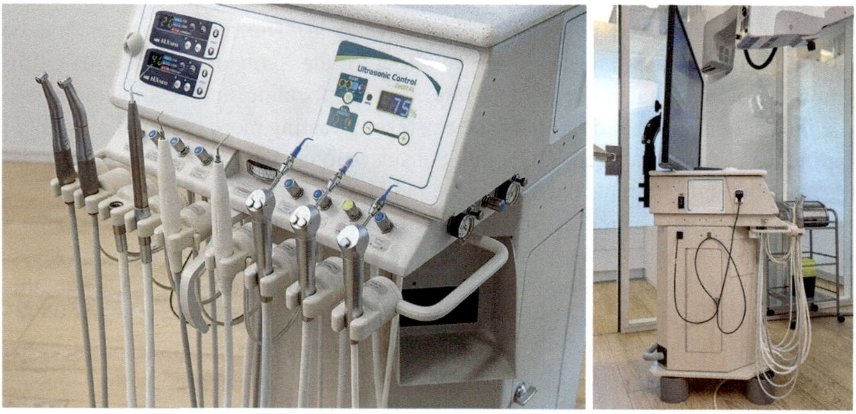

Fig. 2.24 Dental cart showing multiple handpieces, ultrasonics, and Stropko irrigators (left) and a USB port for connecting the digital X-ray sensor (right)

operator to have to elevate the elbow from the armrest, causing strain on the shoulders. This height will be mainly corrected by replacing the wheels of the cart. The larger the wheels, the higher the position of the instrument cradles.

The number of handpieces or ultrasonic handpieces remains an operator-dependent decision. Placement of multiple handpieces or ultrasonics in the cart will facilitate using different burs or tips at the same time without causing interruptions by changing them. Having too many will reduce the space for other instruments or irrigators. Another consideration is if other endodontic equipment like apex locators, rotary handpieces, or obturating devices should be included. In more recent times, many of these devices have had a cordless redesign, so there is no need for them to be included in the cart. Cordless devices allow the operator to use them using no more than class I and II motions (the assistant hands them over to the operator) as opposed to corded ones that will require the use of a class III motion (to grasp them from the cart). Another potential problem of equipping the cart with these devices is the risk of any of them needing repair replacement will increase maintenance and repair costs.

Stropko irrigators are an essential part in the armamentarium of any endodontist and restorative dentist. The number of Stropko irrigators will depend on the treatments performed. A cart set up for microscopic endodontics should be equipped with three different types of irrigators: air/water, reduced pressure air/water, and air-only. The air/water irrigator is mainly used for any routine procedures. The air-only irrigator is used for restorative procedures where an environment free of any water of moisture is mandatory, for example, with bonding procedures. Finally, the reduced air-water irrigator is used for more specific tasks, such us situations where small air puffs are used for inspection or where water is used to rinse canals or washout temporary intracanal medications.

Another component that could be included as an option is a built in compressor. They are small, relatively inexpensive, and require little maintenance, as well as they eliminate the need for a larger compressor at a distant site. This can be an advantage in situations of compressor failure. In a situation where there is one large compressor for multiple operatories, in case of failure, the treatment will have to be stopped or postponed. When carts with built in compressors are used, the treatment can be continued at a different operatory.

The final consideration for a cart is the integration of a computer and monitor. Placing an independent computer attached to the cart has many advantages. It will allow the operator to have full view of the patient file to check any data or treatment details whilst located in the circle of influence. It will help with case presentation, patient education and allow the use of a radiographic sensor. The images will be immediately uploaded onto this computer and will be stored in the patient file. If a ceiling mounted microscope power box is positioned above the cart, enough clearance should be left so the monitor can pass without hitting the microscope power box.

Microsurgery Layout Considerations

An efficient operatory is one where any endodontic procedure, surgical or non-surgical, can be performed at the highest ergonomic level and that can be readily set up within a few minutes. A microsurgical procedure, as opposite to a non-surgical, is more efficient if performed with two assistants (Fig. 2.25). During the procedure, the first assistant and the operator remain at the microscope the whole time, just using class I and II motions. Their position throughout the procedure is fixed and very similar to the already described position for endodontic treatment: 12 o'clock position for the operator and 3 o'clock for the assistant.

Any instruments to be employed during the surgical treatment are brought in and out of the operative field by the second assistant, who sits facing the operator at a 9 o'clock position, performing all tasks using class I, II, and III motions without turning or twisting the back. The second assistant is "the eyes of the operator and first assistant outside the microscope", sitting with the instrument tray on the cart and with a screen positioned directly in front, where the procedure can be followed and anticipated. For these procedures, the cart is turned 180 degrees so the handpieces and ultrasonics can be easily grasped and handed to the operator.

When designing the operatory, enough room has to be planned between the operator and the wall on the right side because during surgical treatment, the cart is moved in a position that allows for the second assistant to sit close to the operator, pushing the cart towards the direction of the wall.

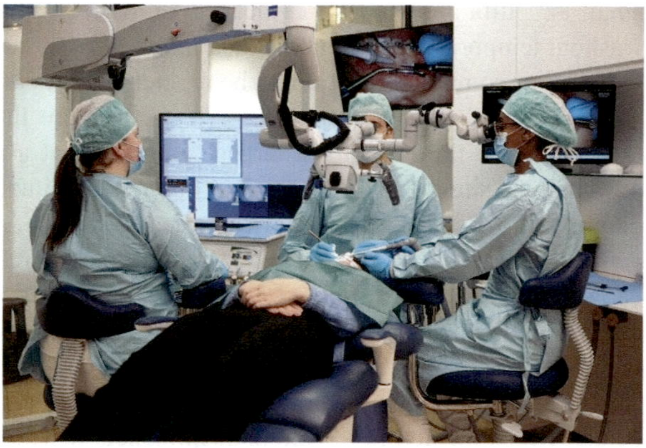

Fig. 2.25 Endodontic surgery performed with two assistants; the operatory should be designed so there is enough room for one assistant to work on each side

Door Locations and Patient/Staff Movements

The location of the operatory doors is directly influenced by the patient, assistant, and operator routes. These routes should be taken into account when deciding the position of the different elements in the operatory. In cases where there is enough flexibility or the doors can be moved, the location should be decided once the orientation of the operatory and the position of the different elements is determined. In an ideal situation, these routes should not cross each other. Single door designs are to be avoided, and ideally three or four doors should be included.

The patient enters the operatory from the waiting room, following an unobstructed route, ideally facing the dental chair, operator, and assistant. The position of this door can be helpful for anxious patients because visualising the door during the treatment can help to reduce anxiety levels (the patient perceives the door unconsciously as an "escape route") (Fig. 2.26).

For the assistant door, this should have direct access to the sterilisation or storage area, to facilitate set up or clearing the operatory after treatment. In the event that another staff member has to retrieve an instrument or material, having direct access to the assistant is advantageous.

The operator route should have a dedicated entry. In an ideal ergonomic setup, the operator would not have to leave the room during treatment. In case the operator has to leave the room, for example, when taking radiographs, a door should be placed in the near vicinity, avoiding having to walk around the chair and the cart.

When the CBCT room is located next to the operatory, a door should also be added to facilitate the CBCT acquisition flow before, during, or after the treatment.

Ideally all entries should have a door that can be closed for privacy reasons. Whenever possible, sliding doors should be considered as they occupy less circulation space. They should be able to be opened and closed easily, ideally using the elbow for hygienic reasons. In some cases, automatic doors could also be considered.

Fig. 2.26 Operator greeting patient, with the exit door positioned in front of the patient

North/South Orientation and Light/Windows

Several studies have shown the importance of natural light in the workplace. Correct levels of natural light can affect several aspects like mood, sleep, vitamin D levels, and eye health. For patients, natural light enhances the production of serotonin, hormone that is directly related to mood, what can be important when treating anxious patients. Windows in the operatory can also represent a problem in the operatory design. Very often these windows are "fixed" and cannot be moved, and this can influence the orientation and the position of some of the operatory components.

Sometimes the presence of windows with a nice view in the operatories can be a determining factor when choosing a location to start up an endodontic practice, but we should not forget that during the treatment the patient lies down, and the operator and assistant are working in the microscope. Also, if these windows are at a ground floor or at any busy street, blinds will have to be used for privacy reasons.

The orientation of these windows will play a fundamental role for natural illumination. In the Northern Hemisphere, windows facing the north are the most ideal solution, as direct sun will not enter directly through the windows. If windows are facing south or in a direct west-east (sunrise or sunset) direction, blinds will have to be used some periods of the day.

As a final remark, if the operatory walls have a large window proportion, it will affect some aspects of the operatory design, like reducing the number of available doors, microscope position (less wall space available, so more dependent on a ceiling mount), umbilical cart position (connections might travel under the floor instead of the wall), or position of any screens (ceiling mounts and important to avoid direct window reflections) (Fig. 2.27).

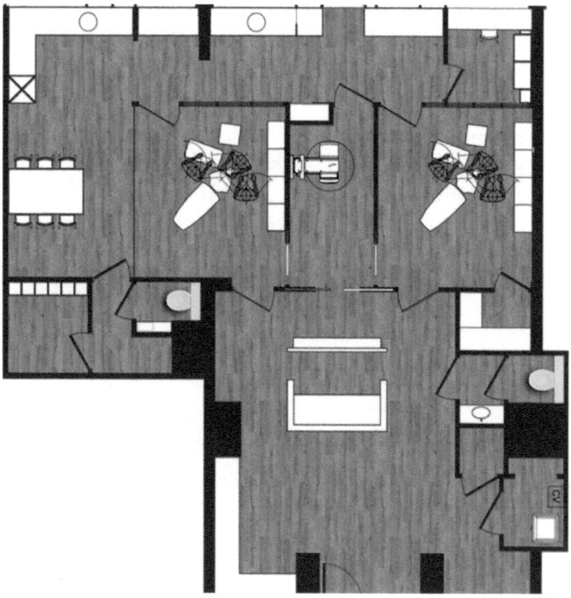

Fig. 2.27 Example plan

Reference

Carr GB, Murgel CA. The use of the operating microscope in endodontics. Dent Clin N Am. 2010;54:191–214.

The Art and Science of Documentation

3

Scott K. Bentkover

Abstract

This chapter describes important photography principles that are relevant to providing the highest level of microscope documentation. The required camera components and equipment are covered. Ergonomic practices teaching teams of operators and assistants how to take top quality microscope photographs are included.

Introduction

The *modus operandi* of microscope documentation has been emphasised in endodontics since the introduction of the Operating Microscope. Although the capture media have changed, early forms of photography and videography were capable of achieving high quality results, at a cost of both a high financial burden and lengthy time commitments. Fortunately, the days of waiting 2 weeks for a roll of 35 mm film to return, or hours spent scrolling through miles of betacam video tape are long gone, but the endgame is still much the same: world class documentation that conveys to the viewer the same level of colour, lighting, sharpness, and depth of field that is seen through the microscope. When Dr. Gary Carr brought the microscope to the forefront of endodontics, he did so through impeccable documentation and was able to convince what initially was an unwelcoming old guard that the microscope was here to stay.

Documentation through a microscope utilises a combination of both art and science. Capturing high quality digital images and video takes place not only with a

The original version of the chapter has been revised. A correction to this chapter can be found at https://doi.org/10.1007/978-3-031-70188-7_9

S. K. Bentkover (✉)
Chicago, IL, USA

K. E. Yapp, C. Aznar Portoles, *Team Microscope Endodontics*, https://doi.org/10.1007/978-3-031-70188-7_3

specific armamentarium of equipment, but through ergonomic practices that elevate ordinary results to extraordinary results. Many authors have cited reasons for the importance of regular digital photography in the dental office, including: diagnosis and treatment planning, medicolegal record keeping, forensic documentation, insurance verification, patient and referring dentist education, marketing, communication with laboratories, dental team members, and colleagues (Ahmad 2004; Terry et al. 2009). The goal of this summary will be to get an individual "up and running" with microscope documentation utilising the current knowledge and equipment available.

Important Photography Basics

Before entering into the discussion of capturing images through a microscope, it is important to have a basic understanding of some fundamental concepts in photography that will be referenced in this chapter. This knowledge will aid the reader in all levels of understanding the equipment necessary for photography, as well as the important concepts behind them. Additionally, these concepts will serve as a foundation to master a thorough knowledge of clinical photographic skills.

The first term that the reader needs to be familiar with is *resolution*. Resolution refers to the sharpness and clarity of an image. It is the ability of an imaging system to resolve detail in the object that is being imaged (CIPA DCG-001-Translation-2005 n.d.; ANSI/I3A IT10.7000-2004 n.d.; Ratcliff et al. 2004). When evaluating the resolution of a digital image, it is common to describe the image in *pixels* or *picture element* (*pix* for "picture" and *el* for "element"), or the smallest addressable element in the image or display device (Foley and Van Dam 1982; Graf 1999). Pixels are typically given as a number of ***megapixels***, which can be calculated by multiplying an image's pixel columns by pixel rows and dividing by 1 million (Foley and Van Dam 1982; Graf 1999) Increasing the number of pixels in an image will more accurately represent the original object (Fig. 3.1).

Another important concept to understand in photography is that of an *aperture*. The aperture is the size of the hole through which light travels through an optical system (Terry et al. 2007). It is important to understand this term as it directly relates to both the amount of light that is allowed into an optical system, as well as the depth of field imparted into images captured through that system.

Depth of field is one of the single most important characteristics of a surgical microscope. The depth of field determines how well it is possible to achieve a focused view of an object within an extended depth (Meditech 2006). It is the distance between the nearest and farthest objects in an image that appear acceptably sharp in an image (Fig. 3.2).

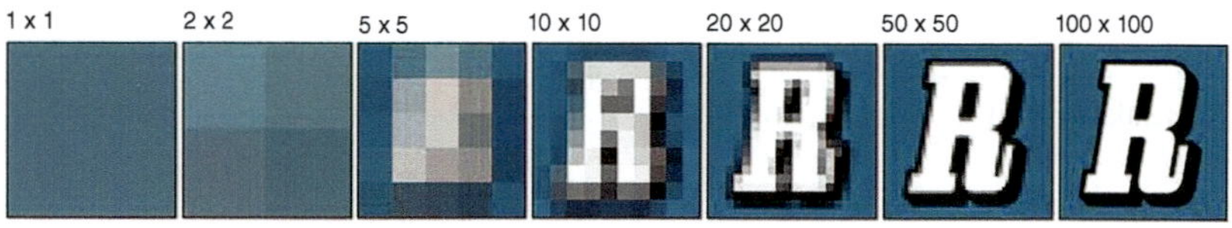

Fig. 3.1 The image becomes more clear as the pixels are increased

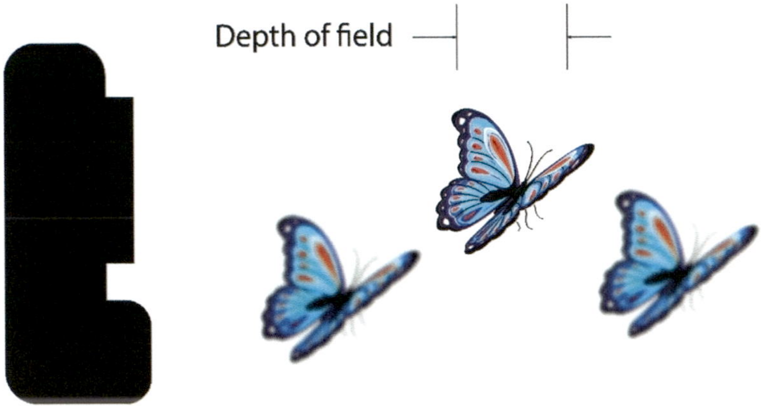

Fig. 3.2 Only the center butterfly is in focus as it is in the depth of field in the optical system

Fig. 3.3 Image taken with a wide open dual iris, therefore less depth of field, and more light

An *iris* can be added to optical systems or lenses to constrict in varying amounts the aperture, which in turn increases or decreases the depth of field (Meditech 2006). Closing down an iris in an optical system will *decrease the aperture* of the system, but will *increase the depth of field* of that system. Also, a by-product of *decreasing the aperture* of a system is to *decrease the amount of light* allowed through the system. Therefore, closing down an iris is the desired method used to increase the depth of field of a system; however, other methods must be used to compensate for the loss of light that incurs such as increasing the power of the light source or increasing the ISO setting of the camera to make it more sensitive (Figs. 3.3, 3.4, 3.5, and 3.6).

Figure 3.5 has greater clarity and detail and is much preferred over Fig. 3.6.

Typically, professional studio photographers desire images with a shallow depth of field to highlight their subjects; however, microscopic endodontics requires the opposite. The root canal environment is deep and narrow, whereby maximum depth of field is desired so as to give observers the greatest amount of detail and understanding. Small versus large depth of field is depicted in Figs. 3.7 and 3.8.

Fig. 3.4 Image taken with a closed dual iris, therefore more depth of field, but less light. This image shows more detail, but will need additional light

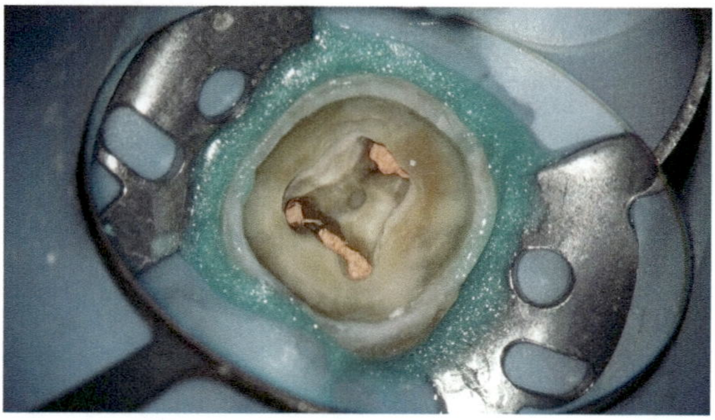

Fig. 3.5 Clinical image with everything in focus, by having a greater depth of field and a closed dual iris

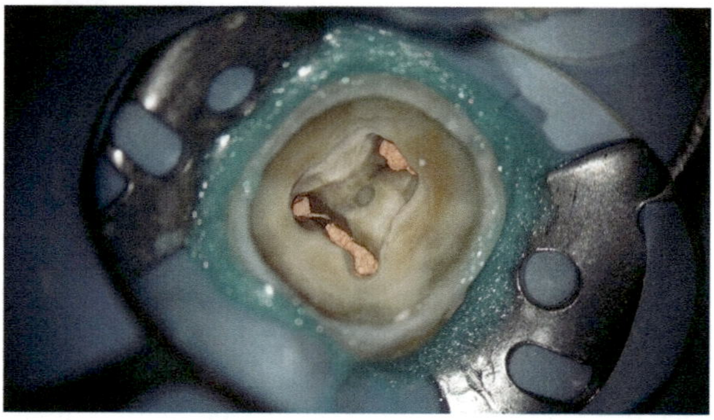

Fig. 3.6 Clinical image with only the gutta percha in focus. The surrounding rubber dam clamp, dam blockout material, and enamel are out of focus

Video link to documentation:

Fig. 3.7 Image taken with a larger aperture, therefore less depth of field (bench only is in focus)

Fig. 3.8 Image taken with a smaller aperture, therefore greater depth of field (bench, grass and tree are in focus)

When specifically analysing the depth of field in an operating microscope, two other aspects of the optical system independently affect depth of field: magnification and the focal length of the objective lens. Increasing the magnification of an optical system decreases the depth of field and can easily be demonstrated when comparing the view of an object at low magnification, to that of the same object at high magnification. An operating microscope might have 45 to 50 mm of depth of field at low magnification, but only 2 to 3 mm of depth of field at high magnification. Additionally, the objective lens focal length will directly affect the depth of field of the optical system. When looking at a basic lens such as an objective lens, focal length is the distance from the plane of the lens to the point that an image is in focus (Note that this should not be confused with the focal length of a photographic lens which will be discussed later). Therefore, a 300 mm objective lens will be in focus further away from a subject and have a *greater* depth of field than a 200 mm objective lens, so *increasing* the focal distance of an objective lens will *increase* the depth of field of the system (Fig. 3.9). A longer focal length also provides a greater working distance and more room for exchanging instruments and materials.

Another important concept in understanding digital photography is that of the ISO, which stands for International Standards Organization, the standardised industry scale for measuring sensitivity to light (Writer 2013). It pertains to the sensitivity of a digital camera's image sensor. ISO is measured in numbers, with the lower numbers being the least sensitive to light, (e.g. ISO 50 or 100), and the higher numbers being the most sensitive to light (e.g. ISO 6400) (Writer 2013). Therefore, a

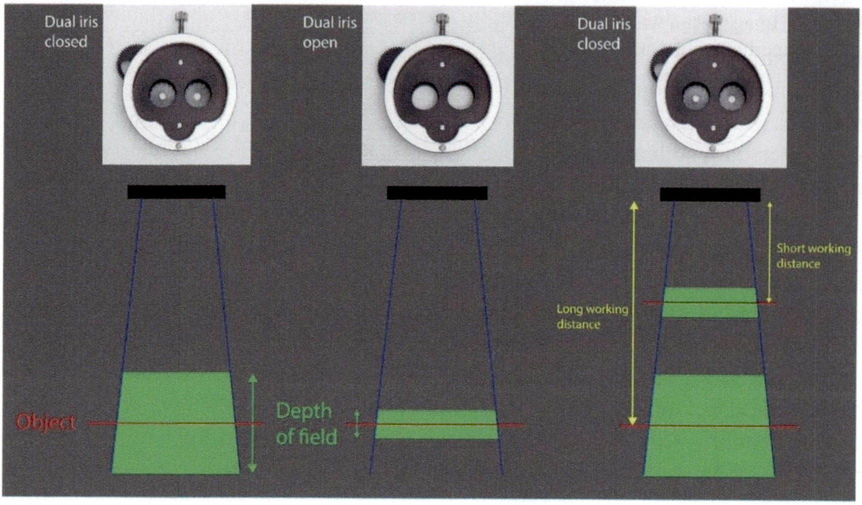

Fig. 3.9 Illustrates how the depth of field range is dependent on the aperture position (left) and the working distance (right)

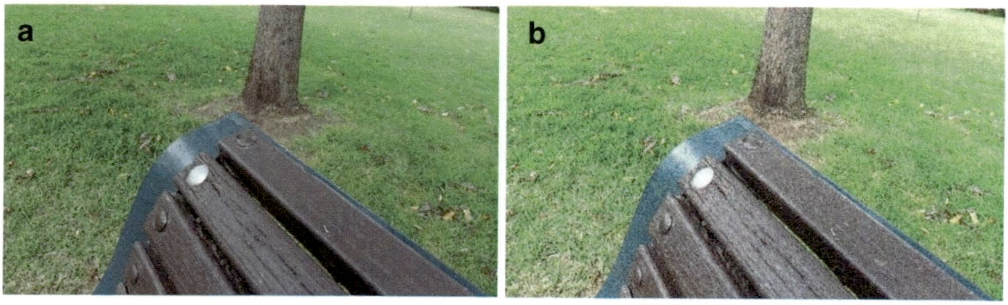

Fig. 3.10 (**a**) ISO 400 (**b**) ISO 3200. Higher ISO gives more light, but allows more noise

camera sensor set at ISO 100 will take more light to produce an acceptable image than the same sensor set at ISO 6400 because the ISO 100 sensor is less sensitive to light. Finally, with increased sensitivity, the camera sensor can capture images in low-light environments without having to use a flash, but higher sensitivity comes at an expense—it adds grain or "noise" to the pictures (Mansurov 2013) (Fig. 3.10).

An important camera function to be concerned with in photography is *shutter speed* or *exposure time,* which is the length of time a camera's shutter is open when taking a photograph (Ray n.d.). This speed is expressed in seconds and is the length of time that your image sensor "sees" the scene you're attempting to capture (Rowse 2006–2012). Longer exposure times will allow the shutter to be open longer which allows more light to be captured by the sensor; however, longer shutter speeds also allow more movement to be captured by the sensor. This might result in blurry images when too much movement is captured by the sensor. The actual speed chosen for a given picture will vary depending upon the operator's ability to control scope and subject movement during the exposure, and how much blur is desired or acceptable. As a rule of thumb, the shutter speed needs to be faster than 1/100 of a second to minimise the effects of scope movement on the image. Faster is better. Higher ISO settings allow for faster shutter speeds but at the potential expense of image noise.

The final basic photography concept to be covered in this chapter that will aid in the understanding photography in the microscope setting is that of *White balance*, which is the process of removing unrealistic colour casts, so that objects which appear white in person are rendered white in your photo (McHugh 2013). White balance is adjusted to get the colours in images as accurate as possible (Rowse 2006–2012). This process can be achieved either manually by using the camera to record a known white sheet of paper, or automatically by having the camera digitally determine the proper white for each image captured. In the past, many of the digital cameras needed the white balance to be input manually. However, fortunately with increased sophistication, most current digital cameras used in microscope photography can detect accurate white balance automatically, eliminating the need for the majority of users to set this through arduous procedures. There will, however, be some discerning users who might see the benefit of performing a custom white balance as it is still the most accurate.

Microscope Armamentarium Important for Photography

When undertaking the task of outfitting a microscope for photography, a thorough understanding of the parts of the basic microscope (e.g. objective lens, magnification changer, binoculars, and eyepieces) will be assumed for the purposes of this chapter. Moving beyond the basic operating microscope setup for capturing digital photography requires the combination of several microscope accessory components. A list of these components can be found in Table 3.1, and are shown in Fig. 3.11.

The first component of particular importance is the *dual iris diaphragm*. As mentioned earlier, an iris closes and opens the aperture, which respectively increases or decreases the depth of field, as well as decreasing and increasing the amount of light the camera sees. An operating microscope has two optical pathways, one for each eye, so each pathway requires an iris, hence the "dual" in the name. Operating

Table 3.1 Complete list of accessories necessary for capturing digital images through a microscope

Microscope components recommended for photography
• Dual iris diaphragm adapter • 50/50 beam splitter • Assistant binocular tube with inclinable binoculars and 10x eyepieces • Photo adapter • DSLR camera • Angled extender and 50 mm spacer for operator binoculars • Operator inclinable binoculars with 10x eyepieces • Reticle for camera side eyepiece

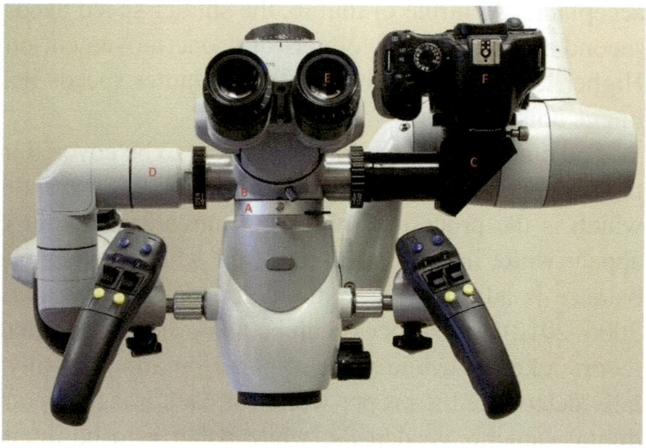

Fig. 3.11 Microscope components necessary for photography. (**a**) Dual iris, (**b**) Beam splitter, (**c**) Photo adapter, (**d**) Assistant binocular tube, (**e**) Reticle (Not visible in photo), (**f**) DSLR Camera

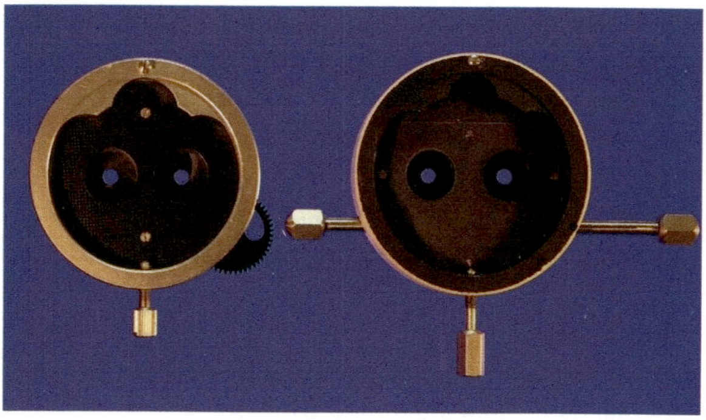

Fig. 3.12 Zeiss Double Iris Diaphragm on the left and the Global Dual Iris Diaphragm on the right

microscopes have inherently poor depth of field, especially at higher magnifications. Adding a dual iris diaphragm to a microscope allows the operator to manually decrease the aperture of the microscope, thereby increasing the depth of field in the system. This invaluable component will impart a greater vertical range of focus in the viewing field, which offers sharpness and more detail in the photos of what the operator is viewing through the microscope. In photography, the equivalent term for measuring the aperture of a camera lens optical system is an "f-stop" or "f-number". However, remember that closing the dual iris diaphragm to increase the depth of field comes along with a cost of decreasing the amount of light passing through the optics. Therefore, the operator might need to make adjustments to the system to increase the amount of light for photography.

Operating the dual iris diaphragm is accomplished either by pushing a sliding bar mechanism to the left or the right (Global Surgical), or by turning a serrated dial (Carl Zeiss) towards or away from the operator (Fig. 3.12).

Either of these two types of motions will either open or close the iris' mechanism. One important distinction to note is that currently, only Zeiss' dial mechanism has a reproducible "click stop" in each position, making the adjustment both reproducible, as well as one that can be accomplished without looking at the instrument. It is advantageous to work at a consistent aperture with a large depth of field. By doing so, this prevents the wasted movement of opening and closing the dual iris for different working and photographical positions. In most situations, the same camera and dual iris settings can be used. If adjustments need to be made to allow more light, ISO setting changes are often the only variable that requires changing. Whilst there is no universal setting for the ideal degree of aperture opening, a rule of thumb would be to select the highest light intensity, then close the dual iris to the point where closing it any further will have insufficient light for clinical procedures. Finally, when assembling the microscope, the dual iris diaphragm is placed directly below the beam splitter (Fig. 3.11). This is so all optical pathways will view the microscope image with the same aperture.

The *Beam Splitter* is another important component utilised in microscope photography. The beam splitter allows the light path to be split allowing a DSLR, video camera, or assistant binocular tube to be attached to the microscope. Typically, beam splitters have two side ports; however, most manufacturers make a single side port version as well, when only one peripheral accessory is desired. The two port beam splitter houses two cube prisms (one for each optical pathway of the microscope) whose function is to split the image light beam so that a portion of the light goes to the side port and the remainder continues on to the eyepiece. The most common beam splitter used is a 50/50 configuration, meaning each prism splits the light in half so that 50% continues on to the eyepieces and 50% goes to the side port (Fig. 3.13).

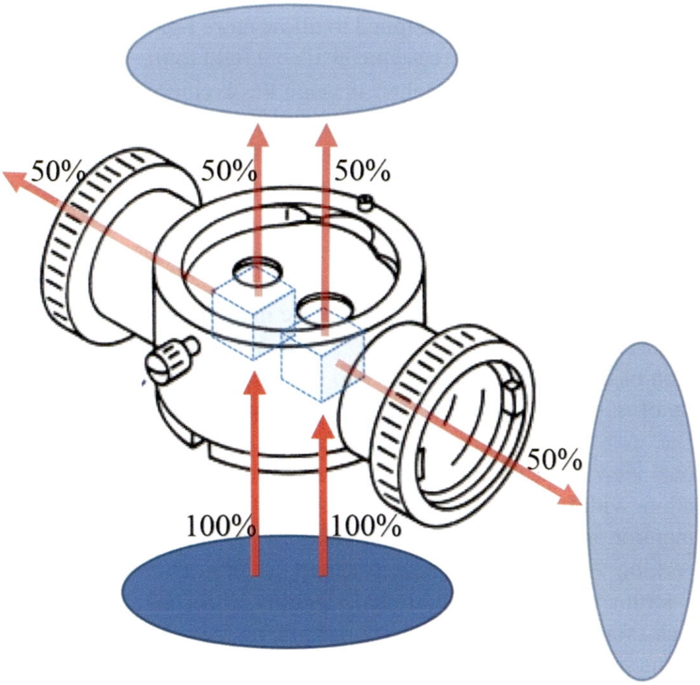

Fig. 3.13 Diagrammatic view of a 50/50 beamsplitter

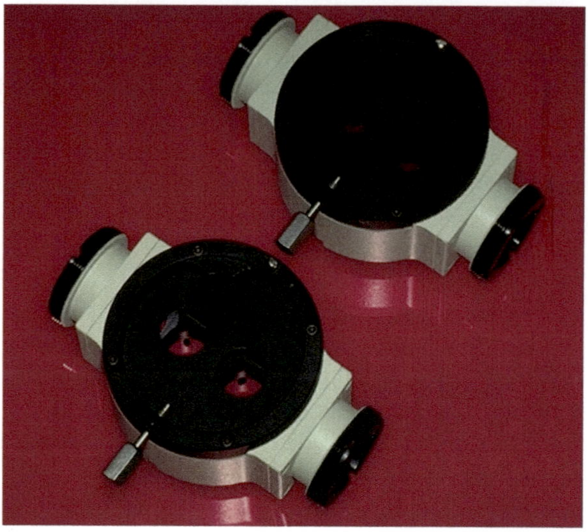

Fig. 3.14 Global Surgical Virtual (bottom left) and 50/50 (top right) beam splitters

Other configurations include the 80/20 beam splitter which sends only 20% of the light to the side port and 80% to the eyepieces, or the Global Surgical Virtual beam splitter (96.5/3.5 split) which sends only 3.5% to the side port and 96.5% to the eyepieces (Fig. 3.14).

These latter two options are designed to allow more light to pass through to the user's eye, but significantly decrease the available light to the side port. This translates into the image appearing brighter through the eyepieces with less light to the camera and was designed to work well with video cameras due to their heightened light sensitivity. DSLR cameras, on the other hand, need more light and do not perform well with the reduced light side port beam splitter configurations because light is a valuable limited commodity in microscope photography and is the biggest obstacle to acquiring great photographs. The 50/50 beam splitter is the only recommended type in modern microscope use.

Perhaps the most understated, and by far the most important microscope accessory is the assistant binocular tube (Fig. 3.11). Aside from the magnification and illumination of the microscope, there is no other component that offers as much to all aspects of micro-endodontics as a trained assistant working through the microscope. The assistant binocular tube offers the assistant an unsurpassed three-dimensional view of the operative site, working in real time and orientation for their own position, with full optical clarity and depth of field that cannot be matched by a video monitor or any other method available on the market today. When an assistant is working through the microscope, they do more than simply follow along, passing instruments or files with limited visibility. Utilising an assistant binocular tube creates a four-handed micro-environment that allows the assistant to be passing and receiving instruments precisely in the location and orientation which the operator needs them, suction directly adjacent to but not in obstruction of the doctor's view, pass materials, and files to the exact location they are to be utilised, perform mirror changes to keep good visibility, remove debris away from the operative site or the operator's mirror to maintain good vision, aid in the administration of anaesthesia, and the list goes on and on. An assistant binocular tube also proves to be invaluable during microscope documentation as well. Assistant documentation skills include: changing to a clean photography mirror, suctioning adjacent to the mirror in all clinical photos to prevent fogging, remotely triggering the camera, changing the camera settings, and organising images.

A key component in the microscope photographic process which connects the camera to the beam splitter is the DSLR photo adapter. The photo adapter is a cylinder with lenses and a mirror or prism that carries the light to the camera so that it focuses directly on the camera sensor and fills it to the maximum level. It acts as a photographic lens that attaches the camera body to the microscope. This small accessory comes fraught with much controversy as there are many criteria which contribute to a superior photo adapter, such as focal length, light transmission, lens quality, image quality, construction, centring ability, weight, and ease of connecting and disconnecting cameras. Suffice to say all are important, however, due to the scope of this textbook, only focal length will be discussed. Focal length of a camera lens is defined as the distance between the lens and the image sensor when the subject is in focus, usually stated in millimetres (Digital SLR camera basics 2013). In photography, longer focal length lenses translate into higher magnification images; conversely, shorter focal length lenses are associated with lower magnification images. The myriad of photo adapters available in the marketplace range in focal length from 170 mm (lower magnification) to 340 mm (higher magnification), which means that some adapters will have twice the magnification of others. Images taken at too low of magnification will not fill the image sensor of a DSLR camera and have a vignetting effect where the image periphery will be darkened and give the appearance of a dark circle around the image. Images taken at higher magnification will fill the image sensor, but as the magnification increases, the depth of field decreases which can lead to a problem where the photo adapter's depth of field is less than that of the microscope. This translates into more out of focus images, as the operator might be seeing an image in focus through the microscope, yet it will be outside of the depth of field of the photo adapter—meaning the camera will see it as out of focus. Also, higher focal length lenses use more lens glass and put the camera body further away from the lens, thereby causing more light loss and producing noticeably darker images in higher focal length adapter. Therefore, at the time of this publication, when considering today's DSLR camera bodies and microscope systems (Table 3.1 and Fig. 3.11), a preferable focal length range would be between 200 mm and 250 mm, however, other adapters would suffice (Fig. 3.15).

The ideal photo adapter has a magnification equal to, or lower than, the operator eyepiece magnification. This will ensure that when properly parfocaled, all images viewed by the operator in focus will translate to camera images taken with the same focus, due to the camera having the same, or greater, depth of field as the operator eyepieces.

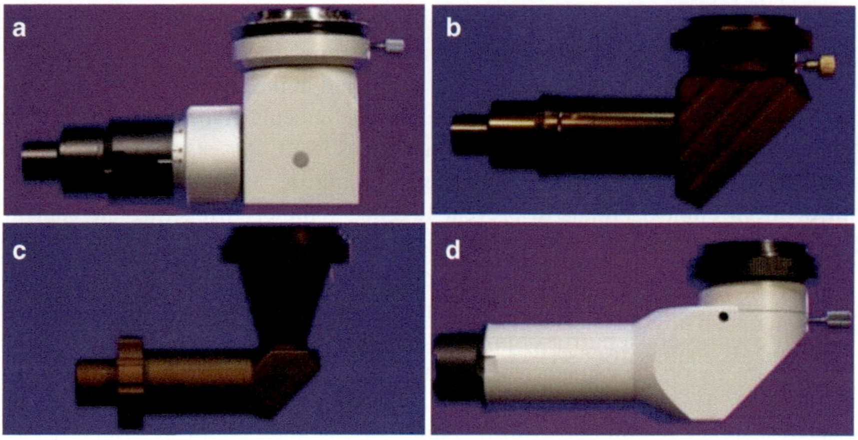

Fig. 3.15 Photo adapters (**a**) Zeiss F = 220, (**b**) DCA 220, (**c**) Global DSLR X-Mount, (**d**) Zeiss F = 340

Another simple component in the microscope photography setup is a reticle, which appears similar to a crosshairs in a gunsight, or a dotted circle and is placed in the eyepiece that coincides with the side of the beam splitter that has the *camera and photo adapter attached to it*. This often forgotten tiny yet indispensable component offers the benefit of allowing the operator to easily centre and frame every image as they are captured. In the industry, there is some confusion as to a common misconception to place the reticle in the eyepiece for the operator's dominant eye. It is the belief of this author that this is incorrect as the operating microscopes utilised in endodontics have stereo optics, which translates into each eye viewing through a separate and independent optical pathway thereby allowing each eye a slightly different perspective. Therefore, the reticle should be placed on the same eye as the camera. If that is not the dominant eye, the operator may find it necessary to close the dominant eye at the time of exposure to insure accurate framing and focus.

One main component of every microscope that directly relates to photography is the light source, where the intensity and colour temperature impart different qualities into the final image. The four types of light sources historically available for microscope use are halogen, metal halide, Light Emitting Diode (LED), and xenon, each offering various advantages and disadvantages. Xenon light sources offer the highest intensity (200kLux), with LED (80–100kLux) and metal halide (50kLux) giving 60–75% less, and halogen giving roughly (50–65kLux) 67–75% less light (all sources vary greatly dependent upon manufacturer and lamp age). These intensity inequalities can mean the difference between needing a ring flash, or capturing images at too slow of a shutter speed with excessively high ISO settings. Xenon lighting also offers the best colour temperature (5400–5500 Kelvin), which is most similar to daylight (5500–6000 Kelvin). Some manufacturer's LED light sources (5500–7500 Kelvin) approximate that of Xenon; however, there is a wide variation in the industry that a blanket statement cannot be made about the colour temperature of LED. But rest assured that LED lighting is the modality of choice for most microscopes when combining the life-span, colour temperature, and price point. Metal halide (7000–7500 Kelvin) and halogen light sources (3500 Kelvin) have a clear disadvantage with colour temperature being skewed away from daylight making them a poor choice as a source of light for photography.

In some instances, microscope photography will need additional light in the form of supplemental macro ring flash source. Microscopes equipped with older halogen, metal halide, and some LED light sources usually do not supply enough light to capture high quality DSLR images when utilising a dual iris for maximum depth of field. In these light-deficient environments, it is beneficial to add a macro ring flash with a guide number (strength) of at least 43′ at ISO 100, such as the Nikon SB-21B or Canon MR-14EX flashes. Operators might also find that the addition of a flash will be necessary, or at least helpful, even with the most powerful Xenon light sources and certain microscope systems. Fortunately, as camera sensor quality continues to improve and allow the operator to acquire images at higher ISO settings with less noise, the need for a flash is less of an issue. With current LED and Xenon light sources having high light levels, the recommendation is to choose one of these light sources so that a flash is unnecessary. This is because there is no advantage or value in using an older light source with an insufficient light intensity.

The final component to be detailed, which sometimes receives more attention than is necessary, is the Digital Single Lens Reflex (DSLR) camera body. A common misbelief is that high quality digital images can only be achieved with the latest and most expensive professional camera body possessing many high end features which increase the cost of the camera. To the contrary, excellent endodontic microscope images are easily achieved with very simple DSLR camera bodies that have been available for many years. When choosing a DSLR camera body, it is important to understand that the majority of DSLR camera bodies that are applicable to our environment are called "crop frame" cameras (or APS-C sized sensors), as opposed to "full frame" cameras (or 35 mm frame-sized sensors) (Harmer 2011). Full frame or larger sensor cameras can have a slight advantage as they will give a cleaner image and will be able to reproduce better detail at higher ISO settings and are therefore much higher priced than their crop frame younger siblings. Fortunately, the crop frame camera bodies do surprisingly well in the ISO range typically found in microscope photography and take equivalent pictures to the full frame sensor cameras, therefore eliminating the need to invest in more expensive bodies. Additionally, crop frame sensors are smaller so they only capture, *or crop*, the centre of the image that a camera lens sends to the sensor. Therefore, the resulting image will have a higher magnification than a full frame sensor camera body with the equivalent lens. This means that the larger frame cameras will produce lower magnification images and will require a higher focal length photo adapter to produce an equivalent magnification image of a crop frame camera, which will negate any advantage the higher quality sensor might offer. Therefore, bigger is not yet better regarding the smaller crop frame cameras. Finally, at the time writing, the "latest and greatest" crop frame camera bodies offer marginally detectable image

Table 3.2 Two methods to achieve Parfocal adjustment. Compiled from a combination of original PERF documents, Zeiss instruction manual and TDO Chat posting by Dr Eric Herbranson

Method 1	Method 2
A. Turn on the power to the microscope B. If you wear eyeglasses, roll back (Global) or dial down (Zeiss) the eye cups to shorten them C. Position the microscope over a dot on a white piece of paper D. Set both eyepiece diopter settings to "0" E. Position the fine focus to the middle of its range. F. Adjust the inter-pupillary distance—"One circle of light" G. Focus the image at the highest power using the fine focus knob *only* to get the clearest image H. Being careful not to physically move the microscope, switch to the lowest power, then focus one eyepiece at a time with one eye closed and turning the diopter ring on the open eye to sharpen the image. Lock into place I. Focus the other eyepiece with the first eye closed and lock the second eyepiece into place. J. Change the magnification to the next highest setting and check each eye for sharpness and adjust if necessary K. The doctor stabilises the microscope, so the assistant can go through the exact same sequence on the co-observer scope to set their diopters L. Repeat several times to confirm the settings M. Record the settings for future use N. Recheck every few months as the operator's eyes change	A. Pull off the binocular head with eyepieces attached B. Set the eyepieces at high - plus(+8) C. Relax the eyes by looking at infinity D. Do one eye at a time E. Look at a very distant subject through the binocular F. Roll the eyepiece through focus until the image is sharp G. Repeat a few times until you have confidence in the settings H. Record the settings for future use I. This procedure should be repeated as the operator's eyes change

quality advantages over the previous generation cameras, so it pays to do a little research prior to purchasing a new camera body for the microscope (Table 3.2).

Getting Started

Once the proper components have been obtained, the final assembly of the microscope can be addressed as depicted in Table 3.1 and Fig. 3.11. During the assembly, care must be taken to balance the microscope arm to accept the additional weight of the components. The next step is to input the settings on the camera as listed in Table 3.3. These settings should be considered a starting point where an operator may alter one or more parameters; however, the majority of users will find that they

Table 3.3 DSLR camera settings

DSLR settings
• Manual • Flash off • TV (Shutter priority) 1/125 • ISO 200 (with flash) or 400–1600 (without flash)[a] • Exposure compensation: None • Image quality to large-fine, medium-fine, or small-fine[b] • White balance—Auto • Metering mode: Centre weighted • Drive mode: Single shot • Picture style: Standard

[a] ISO can vary depending upon microscope manufacturer and light source

[b] Depending upon your image size preference (choose small, medium, or large), but always set the compression to fine

are sufficient. A good rule to keep in mind when taking images through the microscope is to keep the system as simple and reproducible as possible. Additionally, the operator should have the mindset to have the photography equipment and settings ready for high end images on every patient, as attempting to alter images "post-production" through programmes such as Photoshop, Photoplus, or GIMP are extremely time consuming and should not be necessary. Only a minority of individuals might feel it necessary to make minor modifications with these programmes which could be desired for lecturing or publishing. Therefore, operators will find that once their microscope photography setup is complete, capturing great images might require minor alteration of the ISO, light intensity, and dual iris settings.

Following assembly, balancing of the arm and setting up the camera, the next step to consider is how to bring the images from the camera to the computer. Users will find enormous time savings in utilising an active USB extension to connect the camera to the computer to allow continuous image transfer and prevent data loss through utilising programmes such as: DSLR Remote Pro (Breeze Systems), EOS Digital Solutions (Canon USA), Capture NX2 (Nikon USA), or Remote Camera Control/Imaging Edge Desktop (Sony USA). These types of programmes will also remotely trigger the camera in place of a wireless remote. Either option is highly recommended as manually pressing the shutter release button will cause subtle movements of the microscope, which will result in poor image sharpness. Also, another time saver and frustration reducer is to utilise an AC adapter power supply for the camera to eliminate battery and system failure during patient treatment. Additionally, high quality mirrors, such as those made by eie2.com and Zirc are advantageous as they reflect more light and have a higher degree of flatness as compared to typical clinical mirrors, which translate into a brighter and sharper image. Finally, operators should strongly consider utilising office software such as TDO, the most comprehensive endodontic practice management software available, to capture and manage their images.

Aside from the physical setup of the hardware, there is one basic exercise that commands mentioning which is often overlooked and of paramount importance in microscope photography. The parfocal adjustment is the single most important adjustment that can be made with a microscope when utilising a camera (video or DSLR) and/or an assistant binocular tube. Parfocaled objectives assure that all the parts of the microscope, for all magnifications, are focused at the same plane. If the operator is not parfocaled, the camera and the assistant binocular tube will not be in focus when the operator has a sharp image. This is the most common reason for capturing out of focus images through a microscope. The mirror position when undertaking the operator parfocal adjustment should be in the likely, or ideal, photographic position (closer to the tooth)—as opposed to the working position (further from the tooth). There are two methods to achieve parfocal (Table 3.2). Initially, this exercise should be repeated several times a day throughout 1 week until a consistent setting is achieved; however, it might be necessary to repeat every few months to a year as the human eye can change in prescription over time. The eye prescription should be checked regularly to ensure the operator has adequately corrected vision as required. Finally, it is important to understand that a user's eyepiece diopter values will vary from one microscope to another.

Ergonomic Photography Skills

When documenting clinical cases through the microscope, there are several ergonomic techniques that can be applied which will not only make image capture easier, but also improve the final results. All too often, individuals will assemble the right microscope component, purchase the correct camera, set up everything correctly, but then forget these simple tasks, and still struggle at some level.

The first basic ergonomic application is that of proper doctor-patient-assistant positioning. The myriad of clinical operating skills all take place whilst the patient is placed in a flat horizontal position with the operator sitting at the 12:00 position and the assistant sitting at the 3:00 position (Fig. 3.16).

Most of the actual *process* of image acquisition is performed by the assistant where once trained in proper image acquisition skills, will be as instrumental as the doctor. The assistant is the key individual who opens the image acquisition

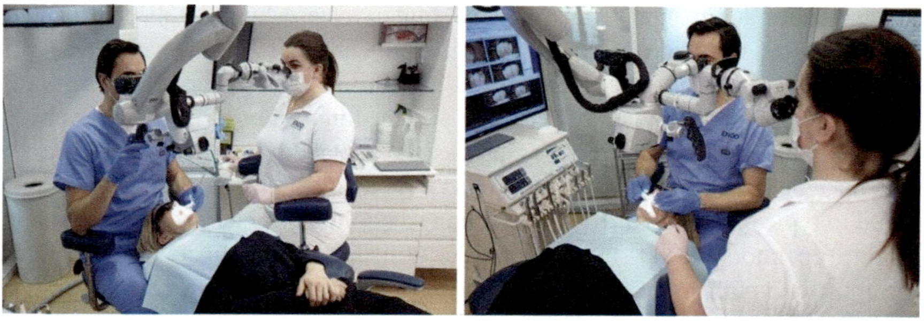

Fig. 3.16 Assistant taking the photo by clicking the mouse (left) and looking at the cart monitor to review image capture (right)

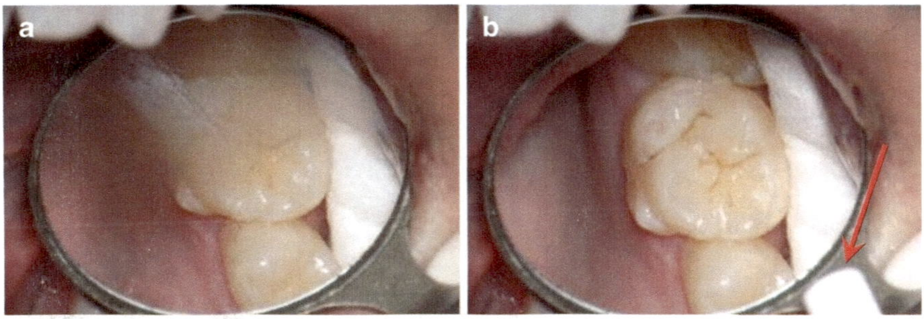

Fig. 3.17 (**a**) Intra-oral photo taken through the microscope with fogged mirror before proper placement of the suction by the assistant. (**b**) Unobstructed placement of the suction easily defogs the large clinical mirror, eliminating the desire to use special solutions or heat the mirror

software, defogs the mirror (Fig. 3.17), triggers the camera, and contemporaneously captures, labels, and integrates the images into the patient's chart.

Thus, whilst applying all the proper microscope clinical techniques, the assistant needs to be able to view the entire procedure through the assistant binocular tube with proper access to the instrument tray and materials, suction, mouse, etc. whilst also being able to see both the assistant and doctor's computer monitors. Through this orientation, it is the assistant, not the doctor, who triggers the cameras through use of either a mouse or remote to take the photo. Image acquisition is never performed by physically pressing the shutter release button on the camera as that creates movement of the microscope and invariably ends up changing the plane of focus leading to areas being not centred or out of focus.

The second important concept of ergonomic acquisition is that of focusing the images. Whilst image acquisition is taking place, the operator is focusing, staging the image scene, and centring the point of emphasis. Once the patient position has been adjusted and photo composition is set, image sharpness should be checked by focusing the objective lens—first by changing to a magnification higher than required for the intended photo. If the microscope has been parfocaled, minimal fine focus adjustment at this point would be required. If the working mirror position is notably different to the photographic position, the body of the microscope may need to be adjusted to have a stable reference and ensure the object is in focus. This is another reason why the parfocal adjustment should occur in an ideal photographic position, so that wasted movement does not occur every time a photo is taken. Once the image is in sharp focus at the higher magnification, the magnification is lowered to ensure the frame is at the correct size for the object. The photo is then ready to be taken by the assistant using the mouse or remote. This process follows a "measure twice, cut once" approach. By checking the focus at a higher magnification, this checks the depth of field of the camera in relation to the subject, and prevents a blurry photo being taken, which wastes both effort and time. Some may suggest focusing the image by moving the patient chair up or down; however, this alters the working position for both operator and assistant. Unfortunately, this would move both forearms into a strained position, either being too high or too low. Inevitably,

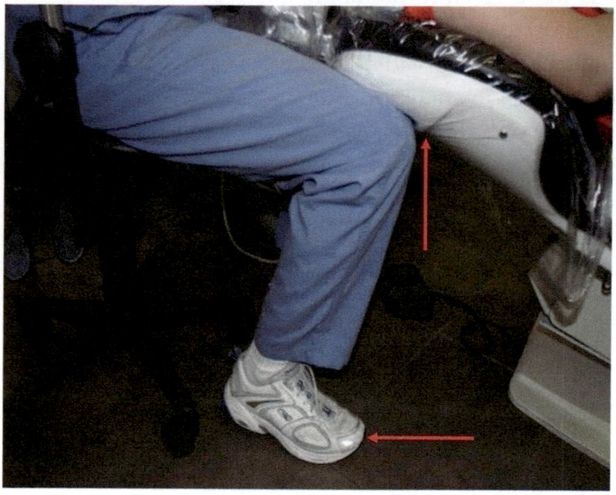

Fig. 3.18 Operator utilising knee and ball of foot to execute precision fine-focusing of the microscope images

the patient chair would then need to be moved back to the ideal working position. This would introduce another wasted movement and therefore it is not recommended to move the patient chair, in order to focus for photography. Any required patient chair height adjustments that differ from the operator's standard preset should be made as part of the starting protocol. However, in some situations, a hands free precision fine focus method is required. This involves not touching the microscope at all, but instead the operator positions the legs under the microscope to use the knee and foot gently, lifting the patient chair. This obtains precision focus without having to move the hands to change the magnification. This can be an advantage during surgery when working with a higher magnification and both the operator's hands are being used (Fig. 3.18).

A third practice that aides the operator in obtaining proper documentation ergonomics is that of mirror type and positioning. As mentioned earlier in this chapter, there are premium mirrors available with ideal coatings to make them more highly reflective, which in turn reflect more light to the cameras. A further nuance of mirror selection comes in the form of both the size of the mirror head, as well as the type of the mirror head. Every practitioner can certainly identify with the cramped working conditions when treating an upper second or third molar, but now consider that this area is no different than any other with image acquisition. Contemporaneous image acquisition makes no exception for the working environment or location. Thus, the added benefit of utilising a 10 mm mirror head over that of a 19 mm or 25 mm mirror head can be the difference between a photo that captures an off-angle portion of the occlusal surface because the working space was limited with a 19 mm mirror head, versus the additional working room offered by a 10 mm mirror head Figs. 3.19 and 3.20.

Furthermore, the type of mirror head can also be beneficial in allowing adequate access and orientation of the head where the flexible mirror heads allow the operator to bend the mirror head into the best orientation surpassing the inability of the traditional fixed mirror heads. Finally, applying a slight rotation of the patient's head

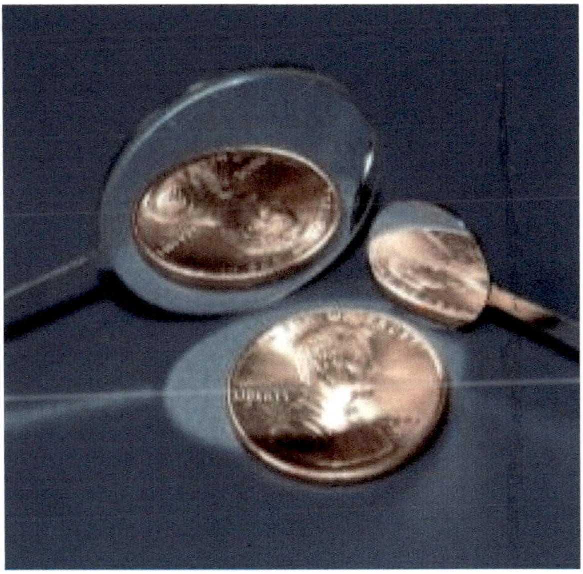

Fig. 3.19 99% Reflectivity of EIE2 mirrors

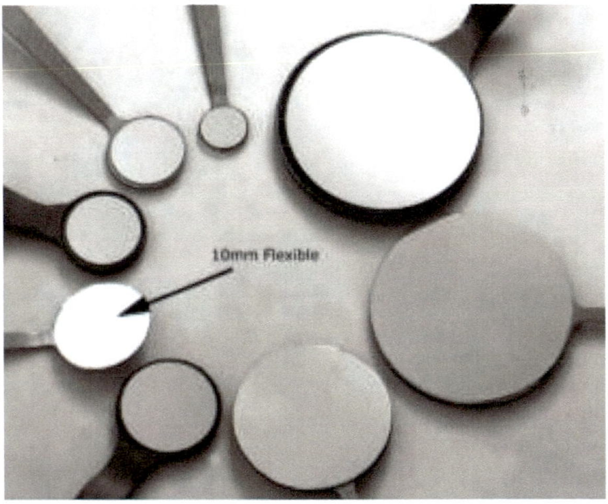

Fig. 3.20 Various sizes of mirrors

as well as bending the mirror head to strive for a 45 degree angle of the reflective mirror surface to the image allows for the most amount of light transmission off the mirror's reflective surface, thereby allowing even more light as compared to an acute angle of the mirror head Figs. 3.21 and 3.22.

A final approach that leads to better efficiency is the method by which the images are viewed by the operator prior to acquiring the images. When surveying the way in which many practitioners utilise microscope photography, it is observed that a common method to view the image prior to acquisition is to look at a separate monitor with a video feed from the camera, to attempt to ensure both adequate focus as well as image centring (Fig. 3.23).

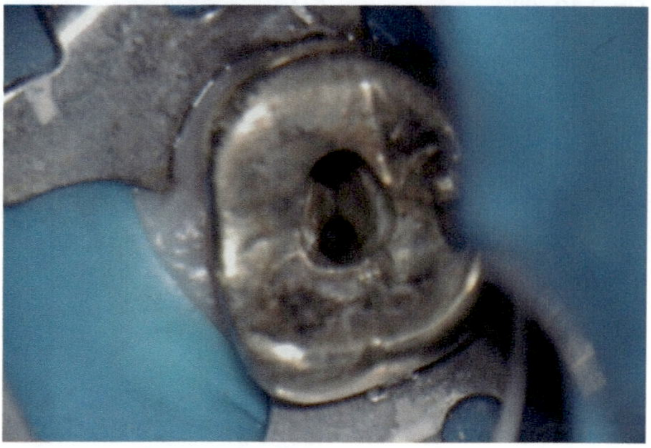

Fig. 3.21 Acute angle of the mirror head causes more of the image to be reflected over a greater portion of the mirror's surface, thereby absorbing more light

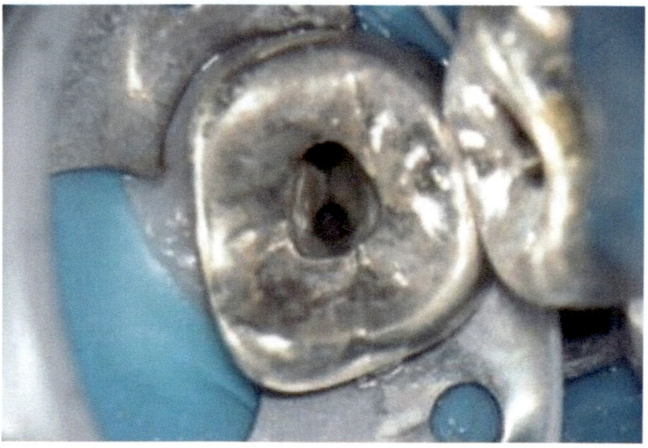

Fig. 3.22 A more ideal 45° angle of the mirror head allow more light to be reflected off of the mirror's surface

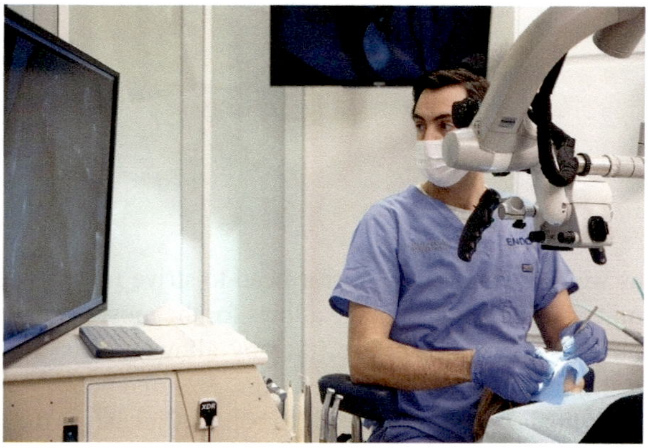

Fig. 3.23 Incorrect ergonomics with wasted movement at the neck and hips when viewing images on a monitor prior to taking the photo

This practice is commonplace because many practitioners have not adequately applied parfocaling techniques, nor utilise a reticle, which results in out of focus images to be frequently seen without the interaction of these two critical concepts. Also, the magnification of the video monitors frequently is lower than what is seen in the acquired images. This can lead to the projected image plane on the monitor being different than the plane at which the image is captured, which will also lead to out-of-focus images. Furthermore, the ergonomics of an operator turning their head to the side to look at a video monitor, versus viewing through the binoculars of the microscope can also lead to a cumbersome working environment. Fortunately, proper parfocaling using a reticle to centre the image, eliminates the need to view images through anywhere other than the microscope—ensuring near perfect, in focus and centred images almost every time.

Putting It All Together

The ideal microscope and photography equipment setup includes: a low enough magnification for the operator to see from under the patient's nose to above the chin, an LED or xenon light source used at the highest intensity setting, an objective lens of approximately 300 mm, a dual iris closed approximately halfway, a 50/50 beam splitter, an assistant binocular tube with angled binoculars and 10x eyepieces, an angled extender and 50 mm spacer for the operator, 10x operator binocular eyepieces with a reticle on the camera side, a photo adapter with a 200-250 mm focal length, and a crop frame DSLR camera.

This setup, with correct camera settings, would allow the operator to take photos without changing any settings in most situations. The likely scenario where any setting(s) would require adjustment would be when taking a photograph of a low-light setting, such as deep inside a canal. In this case, the assistant would increase the ISO setting, making the camera sensor more light sensitive. This system is as simple and reproducible as possible. Consistency is an important feature in training your skillset, where controlling the multiple photographic variables allow you to isolate them, identify any shortcomings, and correct them immediately—leading to perfectly taken photos.

Limitations of using older microscope equipment can be significant, to the point where attaining world class documentation is impossible. A common scenario is having *an insufficient light source*. Viewing an image through the microscope may have an acceptable light level; however, the photo being taken will be too dark and lack any relevant detail, called *underexposure* in photographic terms. The first variable to change would be to open the dual iris and allow more light to the camera. Whilst light exposure is increased, depth of field is decreased and certain parts of the foreground and background will be out of focus. Another variable is increasing the ISO setting. This increases light exposure, at the expense of increasing image graininess or noise. Attaching a ring flash as an adjunct light source will introduce issues of fit, weight, bulkiness, an additional light exposure variable and possibly unwanted shadows in the photo. Whilst limitations in equipment quality can be addressed with compromise, they invariably lead to lower quality photographs. These compromises introduce wasted movement and inefficiencies.

A step-by-step guide for taking high quality microscope photographs is as follows:

1. Check all microscope and camera settings as above.
2. Change to an eie2.com or Zirc mirror that is without scratches or surface damage.
3. Compose the photo in your mind. Pick which part of the subject is going to be the emphasis and ensure it is centred, or close to the centre of the image, using the eye with the camera attachment.
4. Position the patient and mirror so the most amount of light is reflected (likely a 45 degree angle to the long axis of the microscope light beam).
5. Increase the magnification to a level much higher than the intended photo composition frame.
6. Change the fine focus so the point of emphasis is in sharp focus. If the microscope needs to be moved, your hand is already on the microscope body and this can be done with negligible effort.
7. Decrease the magnification to the ideal composition—likely the same as determined in step (4).
8. If an assistant suction or air/water syringe is required to defog the mirror, ensure this is in place and that everything (mirror, patient, assistant) is completely still.
9. Have the assistant remotely trigger the camera and take the photo.
10. Move your eyes to check the monitor and if needed, change (increase) ISO setting if shooting in a low-light scenario.

If there is an issue with photo sharpness, recheck the basic principles of eye prescription, diopter settings, parfocaling, and dual iris setting to ensure the operator's depth of field translates to a similar one of the camera.

Video

The principles of microscope videography are identical to that of microscope photography. As discussed earlier, time spent viewing and editing video in "post-production" is time consuming. There is often greater value in a set of high quality clinical photos, than a passage of video, for communicating a clinical scenario. This section would mainly be for those who use short video clips for teaching and publishing.

Modern DSLR and mirrorless cameras have the ability to record video in addition to taking photos. Given that good quality video documentation is available in lower end cameras, this can be a valuable addition for no additional equipment required. The factors to consider when recording treatment procedures are: file duration and size, battery life, and overheating.

European Union laws, at time of writing, have an import duty on video cameras—Parliamentary question—E-001277/2016(ASW). If a camera can record video of at least 30 min, it is deemed a video camera and eligible for import duty. For this reason, many camera manufacturers only allow recording for up to 29 min and 59 s. Video recording directly through the camera will therefore be limited to under 30 min, before having to restart recording. If recording high resolution 4 K video, file sizes can be large and memory card capacity will be a potential limitation.

Whilst the camera is recording video, battery power will be used and heat is generated. Both of these factors will contribute to the unit automatically powering off. Even with an external power supply, the prolonged heat will be an issue, given that there is no external cooling system for the camera.

The way to overcome these issues, in addition to providing a live video feed for your second assistant during surgery, is to live stream from the camera and record the output using an external recorder. More importantly, *video streaming and recording should be done whilst being able to take photos without interruption.*

High-Definition Multimedia Interface (HDMI) connections are how uncompressed video data are transmitted. The HDMI output on the camera is how high-definition video is streamed to a monitor. To send the signal to a monitor and an external recorder, an HDMI splitter is required, which has one HDMI input and at least two HDMI output ports. One of these outputs should be for the second assistant's monitor for surgery. The other output should be connected to an external recorder (Magewell USB Capture HDMI Plus), which has an input and output and the feed can be displayed to a second monitor. This recorder is connected to the assistant's computer, who will be able to start and stop video recording with a mouse click. An external hard drive is connected to the recorder ensuring there is plenty of space for the large file size of high resolution video. An external power supply should be used, so the camera can be used all day for live video without interruption, due to overheating or lack of power. Currently, the cameras known to be able to perform all of the above are the Sony α6000 series. This is a cropped frame sensor (APS-C) mirrorless camera range, which is compatible with an f = 220 mm photo adapter.

Video composition, with understanding of angles, lighting and aperture settings are implemented in the same way as microscope photography. If the basics of taking excellent photos are used, this can translate to recording good quality video. For entire treatment procedures, capturing an external video feed from the camera is the most reliable method. Short sequences can be taken using the onboard video. One of the most important factors in choosing to record microscope video is to determine what it is used for (teaching, publishing) and understanding how much editing time in "post-production" is required—usually several hours. Often a single photo, or series of photos, can communicate much more value than a short video clip. This is usually because a video has multiple points of attention, making it difficult to highlight the point(s) of emphasis. By controlling the attention of your audience with a photo, the intended message can be clearly shown to the viewer.

Conclusion

Mastering good clinical photographic skills should be a goal for every clinician. This summary can and should be used as a laundry list of how to get started utilising the current knowledge and equipment available. The doctor and assistant will find that taking superior clinical photos and communicating with images will be one of the most rewarding aspects of clinical practice.

References

Ahmad I. Digital and conventional dental photography: a practical clinical manual. Chicago: Quintessence; 2004.

ANSI/I3A IT10.7000-2004 (n.d.) Photography—digital still cameras—guidelines for reporting pixel-related specifications

Bengel W. Mastering digital dental photography. 2nd ed. Chicago, IL: Quintessence; 2006.

CIPA DCG-001-Translation-2005 (n.d.) Guideline for noting digital camera specifications in catalogs

Nikon. Digital SLR camera basics. Nikon Corporation; 2013.

Foley JD, Van Dam A. Fundamentals of interactive computer graphics. Reading, MA: Addison-Wesley; 1982. ISBN 0201144689

Graf RF. Modern dictionary of electronics. Oxford: Newnes; 1999. p. 569. ISBN 0-7506-4331-5

Harmer J. The difference between full-frame and crop sensor DSLR cameras. Improve Photography. 2011;

Mansurov N. Understanding ISO—a beginner's guide. Photography Life. Photographylife.com. 2013;

McHugh S. Tutorials: white balance. In: Understanding white balance. Cambridge in Colour; 2013.

Meditech CZ. Depth of field OPMI® application tip #2. Informed. 2006.: Issue #2:14–7.

Ratcliff S, Bush DE, Fondriest JF. Digital dental photography: a Clinician's guide. Key Biscayne, FL: L. D. Pankey Dental Foundation; 2004. p. 2–3.

Ray SF. Camera features. In: Jacobson RE, et al., editors. Manual of photography: a textbook of photographic and digital imaging. Focal Press; n.d. p. 131–2.

Rowse D. Introduction to Shutter speed in digital photography. In: Digital Photography School RSS. Digital Photography School; 2006–2012.

Rowse D. Introduction to white balance. In: Digital Photography School RSS. Digital Photography School; 2006–2012.

Terry DA, Snow SR, McLaren EA. Contemporary dental photography: selection and application. Funct Esth Rest Dent. 2007;1(1):37–46.

Terry DA, Snow SR, McLaren EA. Communicating digitally with the laboratory: Design, impressions, shade, and the digital laboratory slip. Inside Dent. 2009:62–7.

Writer BH. What does ISO stand for in photography and why do i need to know? Bright Hub. 2013;

Training Principles

4

Abstract

Designing a team training programme that develops expertise involves a deeper understanding about why you are training, and what bigger picture goals your team wants to achieve. Strategies and methods on how to develop expertise are covered. The type of mindset and ways for the team to self assess learning progress are discussed.

Key Points

- The most important principle when a team trains together is to develop trust between all members.
- Training should help you develop ways of thinking and adaptation that develop expertise.
- Receiving specific feedback, identification of errors, and the immediate opportunity to respond are the ways for a novice student to improve and become an expert.

What does it truly mean to be an expert? The Merriam-Webster dictionary defines an expert as "having, involving, or displaying special skill or knowledge derived from training or experience". The Cambridge Handbook of Expertise and Expert Performance describes an expert as "someone widely recognized as a reliable source of knowledge, technique, or skill whose judgment is accorded authority and status by the public or his or her peers" (Ericsson 2006). In simple terms, it is someone who thinks and operates at the highest level of performance possible. More

K. E. Yapp, C. Aznar Portoles, *Team Microscope Endodontics*,
https://doi.org/10.1007/978-3-031-70188-7_4

importantly, expertise—the characteristics, skills, and knowledge that distinguish experts from novices and less experienced people—can be developed. Learning to be an expert follows a simple, but not easy, process.

The fundamental aim of this book is not to teach you a set of movements, nor to give you a recipe for performing a procedure. The goal is to help you learn to develop ways of thinking that transform you and your team into experts in all microscope procedures. This only occurs with a specific, dedicated type of training, accompanied with a mindset that grows and adapts to new challenges.

Begin with the end in mind is a saying with an emphasis on understanding the end result, the journey of your intended direction from your origin and most importantly, to envision what is obtainable, but not currently there. This is so you can plan your path of what is possible. The end goal is to have learned a new language—team microscope endodontics—that has almost no overlap to any other skill, whether it be "naked eye" dentistry or otherwise. Before this new skill is acquired, the most important stage of learning is to embrace the fact that using the microscope as team is a new language and all students start at the very beginning, literally from day one. Regardless of the number of years of experience as a dentist or assistant, it has no impact on how this brand new skill is learned.

Two types of mindset have been described by Carol Dweck, author of *Mindset: The New Psychology of Success* (Dweck 2008)—a growth and fixed mindset. In order to become an expert, you need a growth mindset. This is because adaptation and improvement forms part of the process of developing expertise. The growth mindset embraces challenges, persists in the face of obstacles, sees effort as the path to mastery, learns from criticism, and finds lessons and inspiration in the success of others. A fixed mindset is largely the opposite and prevents skill development or improvement in ability. When you have a fixed mindset, challenges are avoided, obstacles are given up on easily, effort is seen as fruitless, criticism and useful negative feedback is ignored, and you feel threatened by the success of others. The message to readers is clear: have a growth mindset and you will reap all of the rewards in the path to developing expertise.

When you train, it is not a simple matter of repeating the same tasks over again. Repetition helps with memory development, but not necessarily skillset development. If you repeat the same mistakes over and over, they are still mistakes. When you get it right, repetition helps with muscle memory and storing patterns for recollection later.

Experts in all fields, from chess players, musicians, professional athletes, and surgeons, have all reached the highest level of performance in their respective domains due to one common reason. They have trained in a specific manner which is "focused, programmatic, carried out over extended periods of time, guided by conscious performance monitoring, evaluated by analyses of level of expertise reached, identification of errors, and procedures directed at eliminating errors." (Horn and Masunaga 2006). This has been called deliberate practice by K Anders Ericsson.

In simple terms, it is practice involving activities specifically designed to improve the current level of performance, under the guidance of a coach. During practice, appropriate and immediate feedback about performance is obtained—given by the teacher and, in later stages of expertise development, using the student's own self-appraisal. This helps the student be aware of the standards of expertise, identify and correct errors, set new goals, and overcome weaknesses. During deliberate practice, failures occur which should be seen by the astute student as (1) a discovery of the right answer and (2) opportunities to improve. In this process, "expert performers continue to strive to attain more refined mental representations, which provide increased ability to control performance" (Ericsson and Harwell 2019).

All of this requires considerable effort, significant and focused mental energy. In the book, *Talent is Overrated*, author Geoff Colvin mentions that the chief constraint of deliberate practice is mental; the required concentration is so intense it's exhausting (Colvin 2016).

Deliberate practice is also designed to develop effective mental representations (Ericsson and Pool 2016) (p. 97), figure out where weaknesses are and focus on getting better in those areas with different methods (Ericsson and Pool 2016) (p. 181).

Whilst there are other practice activities where the student actively engages to improve their performance without a teacher, termed *purposeful practice* (Ericsson and Pool 2016) (p. 34), the concept of attaining a performance improvement regarding a specified goal with a form of immediate feedback followed by an opportunity to improve, still applies. It is likely that a student will have occasional sessions with a teacher or coach or seek feedback from peers.

Trust. This is the most important component of a team. No matter the size of the team, difficulty of the required tasks or amount of responsibility, trust determines how the team performs at a group and individual level. It defines the team's success and dictates the most relevant outcome—how well it performs its duties (Fig. 4.1).

The basis of trust rests on communication. Because the entire patient procedure focuses on using the microscope, the most efficient way for the operator and assistant to communicate is to do so without looking out of the microscope or interrupting the focus on the task. This primarily involves non-verbal communication, where hand signals and motions are used to initiate leads and responses for all movements. Due to the limited verbal and visual communication, the sense of touch is the most critical in developing patterns to be stored in muscle memory. Hand movement without looking directly at the hands occurs for almost every microscope procedure; this is probably the biggest difference between microscope and naked eye dentistry. This is why all movements occurring outside of the microscope field of view strongly rely on pressure perception in not just exchanging instruments and materials between assistant and operator, but also preventing any instruments from hitting, or being dropped on, the patient. The operator and assistant have to be able to trust

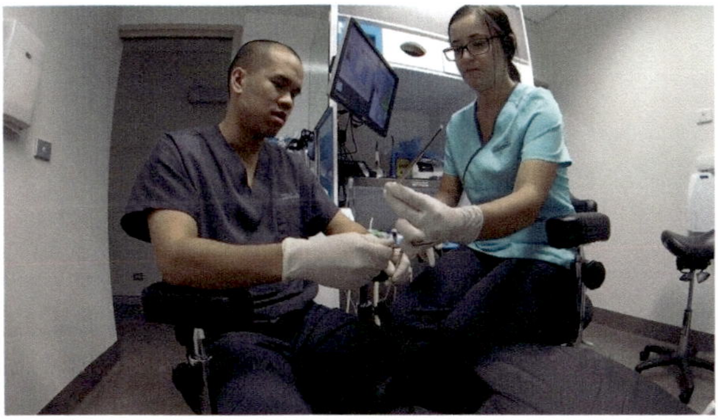

Fig. 4.1 Training without a patient to develop trust between team members

each other to protect the patient from any dangerous situations involving sharp instruments or other materials.

An element of non-verbal communication is the lead and response. The stronger the lead, the clearer the message—meaning, the more likely an appropriate response can occur. Highly functioning teams have a strong leader who clearly lead, and other team members who can clearly respond to the lead and follow orders. This is seen in many domains such as in the military, performing arts and professional sports teams. There must be a lead for a response to occur. Without the lead there is no response. The response can only occur after the appropriate lead has been shown. This response also has to be clear and strong in order to communicate that the lead has been properly received. In microscope endodontics, a strong lead would include an operator's signal to receive, or the "1–2" signal to pass an instrument back to the assistant (Fig. 4.2).

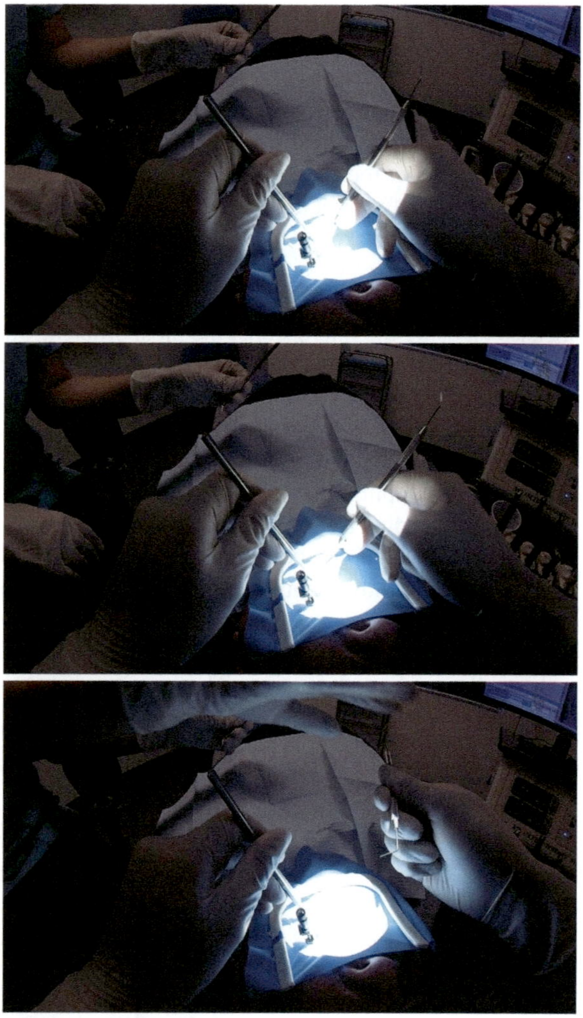

Fig. 4.2 The 1–2 signal to pass an instrument. Hand at rest. "1"—slight upward movement. "2"—slight forward movement

In partner dancing such as Salsa or Bachata, the leader properly leading and the follower responding appropriately is one of the most critical components in a successful performance—this is no different to using the microscope as a team.

Two main types of learning are used in skillset development—declarative and procedural learning. Declarative learning consists of facts (declarations), such as "tooth 46 is a lower right first molar". There are two types: recognition and recall. An example of recognition is showing someone an instrument, who then states its name and what it is used for. Recall is asking someone to list the instruments used to place rubber dam, as another example. It is easier to develop recognition than recall memory; recognition can be considered "primed recall". Declarative memory increases in strength with use and decreases with lack of use. The best way to understand declarative learning is "use it or lose it".

Procedural learning is learning how to carry out a procedure and undertake the required actions in a specified order. Procedural memories are created after declarative memories are available. Without proper declarative learning, failure of procedural learning occurs. This is why declarative learning is so important in developing expertise. Between these two types of learning, there is an intermediate stage—mixed learning. This is when declarative learning is consolidated into procedural learning (Table 4.1).

Mise en place is a French culinary term which means "putting in place". In a kitchen setting, it refers to having all ingredients ready and prepared prior to cooking. Whilst this describes the setup phase of the cooking procedure, the underlying focus is to plan the entire process, start to finish, and anticipate all the required steps, utensils, and ingredients. It is no different to any dental procedure, especially in team microscope endodontics. This is the first part of skillset development—understanding what we do, why and how we do it, what is used and when we use it. Simply put, it is preparing for the recipe before following it (Fig. 4.3). You have to understand the entire procedure before undertaking your setup, or *mise en place*.

To begin, a written list of every instrument and material for the procedure is needed. This reinforces declarative learning and ensures that no item is forgotten. A list of every procedural step should follow. Atul Gawande in *The Checklist Manifesto* (Gawande 2010) writes: checklists in a healthcare setting "are quick and simple tools aimed to buttress the skills of expert professionals. And by remaining swift and usable and resolutely modest, they are saving thousands upon thousands of lives". Checklists also "seem able to defend anyone, even the experienced, against

Table 4.1 Different types of learning

Declarative learning	• Learning facts • Recognition: "What is this instrument/tooth number/surface?" • Recall: "Take out a DG 16 explorer"
Mixed learning	"Orientate the rubber dam, frame, forceps and clamp in the right position in relation to the patient"
Procedural learning	"Perform a one handed instrument exchange"

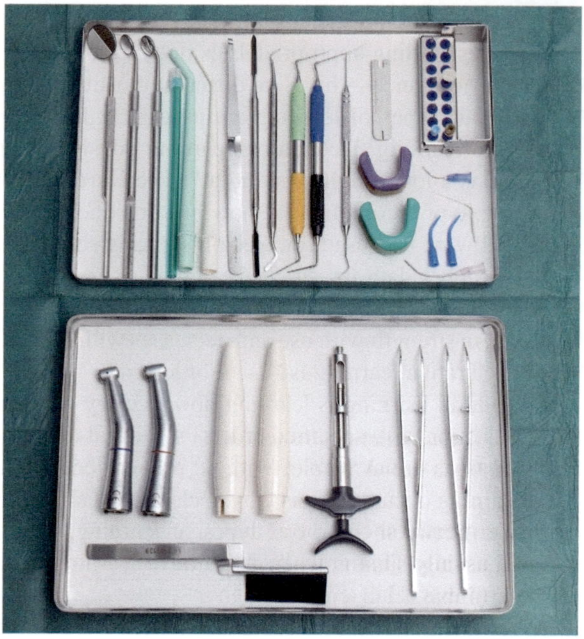

Fig. 4.3 Instrument setup prior to treatment

failure in many more tasks than we realized". Most procedures, especially in endodontics, follow an identical and consistent sequence a majority of the time. Sometimes slight variations occur, but the bulk of the time often defaults to the main procedural order. New operators and assistants should refer to this list so that they can have an understanding of the entire procedure before training commences.

Movement Principles

The main principle to follow is that all procedures should occur with the minimum amount of wasted movement. This is done during training and when the patient is in the chair. If wasted movement is to be limited when working in the operating field, the operator's hands should stay in the field or as close to it, where possible. Otherwise, there will be a lot of wasted time and energy in constantly moving to and from the operating field.

Understanding the classes of movements will give you a clearer insight on what types of motions are actually occurring when you are operating with a patient in the chair. They were described in Chap. 2 and are:

Class 1 - movement of the fingers (hand filing—watch winding)
Class 2 - movement of the wrist (packing gutta percha with condenser)
Class 3 - movement of the elbow (taking the handpiece from the dental cart)
Class 4 - movement of the shoulder (changing microscope magnification)
Class 5 - movement of the hips (leaning to take an instrument from the tray)

Where possible, we try to limit movements in the operating field to class 1 or 2 movements. This should be a major focus when training. If we have to use a class 3 movement like taking a handpiece, the aim is to use a class 3 and not a class 4 or 5 movement. As described in the operatory design section in Chap. 2, class 1 or 2 movements can only be done if the operatory design and equipment allows it—specifically if the cart is positioned in the circle of influence and the operator chair has armrests.

Sometimes a class 4 movement is inevitable (e.g. changing microscope magnification); fortunately these are uncommon and form a minority component of treatment procedures. The team approach should also focus on these movement principles, such as having instruments and materials exchanged next to the operating field, instead of the operator leaning and reaching for an instrument.

Another key component is to keep the microscope still. Do not move the microscope, move the patient (Fig. 4.4).

This allows the operator to see all areas and angles of the mouth with a class 2 movement (turning the patient's head with the wrist) instead of inefficiently moving the microscope using a class 4 shoulder movement and spending more time in doing so. Choosing the right microscope makes this possible—also described in Chap. 2. The microscope field of view has to be large enough to visualise the patient's nose to chin. This emphasises the value of having low magnification. The only way to do this and keep the mouth in the field of view is to swivel the patient chair at the same time by using your knees or legs. Without a patient chair with a swivel function,

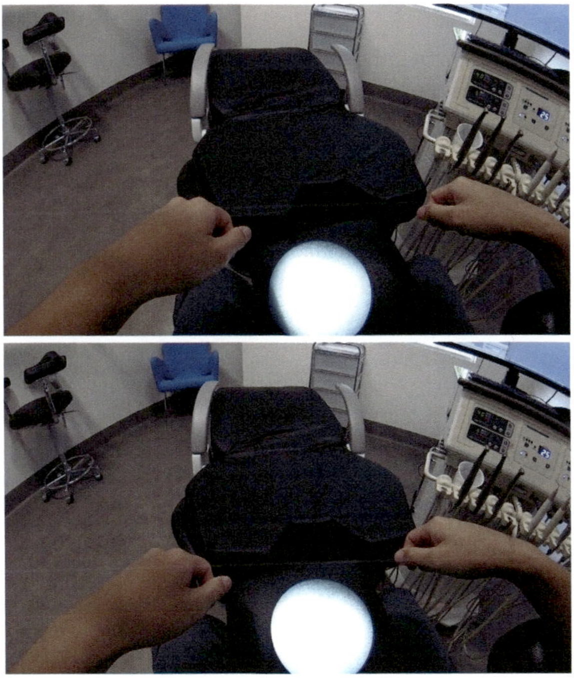

Fig. 4.4 Swivelling patient chair with the legs, which moves the patient using minor adjustments

these class 4 movements cannot be prevented. Training should keep this in mind, where good microscope habits are developed early.

Swivelling the patient is so important because the operator's hands require a certain position or angle, relative to the tooth, to undertake that part of the procedure. An example of local anaesthesia illustrates this point well. A right-handed operator administering a buccal infiltration of anaesthetic for an upper left molar, or palatal infiltration of a right maxillary region, requires the left hand to retract the cheek. The right hand then needs to cross over to the patient's left, so the anaesthetic syringe can reach the target site and not be hindered by the patient's front teeth. This causes the operator to raise the shoulder and lift the right elbow off the armrest, a class 4 motion (Fig. 4.5).

To minimise these wasted movements, instead the patient's head is turned to the right, the patient chair is swivelled to the left, keeping the mouth in the field of view (Fig. 4.6). The anaesthetic can be administered with the operator's right elbow on the armrest with the right shoulder being still. This prevents the unnecessary raising of the shoulder and makes the procedure more efficient by not wasting time and movement.

By following these ideas of minimising wasted movements and keeping motions to either class 1 or 2 where possible, this leads to team members operating together at the highest level of safety and efficiency.

Everything complex is done before you enter the room. This is a training strategy used in many domains, which allows you to prepare mentally and physically for the procedure before it actually occurs. It also gives students an opportunity to anticipate any adverse or unexpected events, which can be addressed with prepared adaptation.

Most importantly, team training with the microscope must first be done without patients. It is unsafe, unprofessional, and very unproductive to try procedures "for the first time" on a patient. Just like in dental school when you did practice fillings and endodontic treatments on extracted teeth before attempting them clinically, the procedure should be rehearsed prior to having a patient in the chair (Fig. 4.7). The difficult parts of the treatment will already be learned and any errors corrected beforehand.

"Bigger picture" training basics of operator and assistant posture, patient positioning, chair, cart, and microscope setup are the first to be learned. This is because these

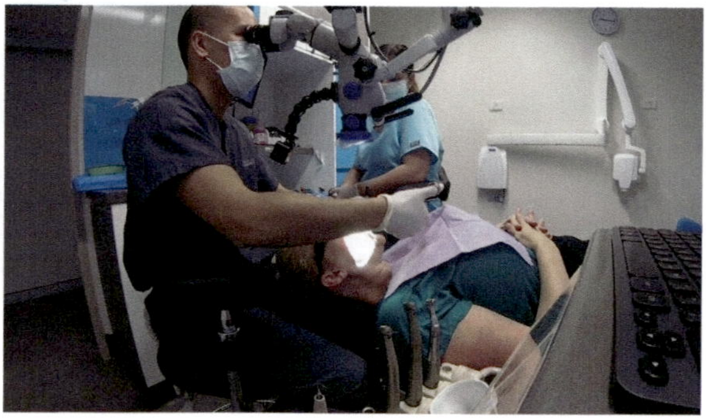

Fig. 4.5 Raising arm and shoulder to reach patient's left side leads to unnecessary strain

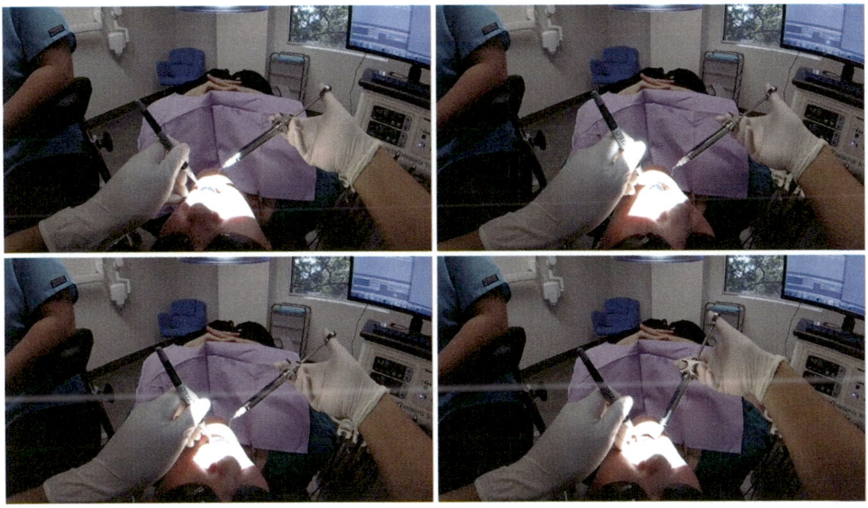

Fig. 4.6 Solution to keep arm on armrest. (**a**, **b**) turn patient's head to the right (**c**) swivel chair in opposite direction to the left (**d**) administer local anaesthetic with arm on armrest

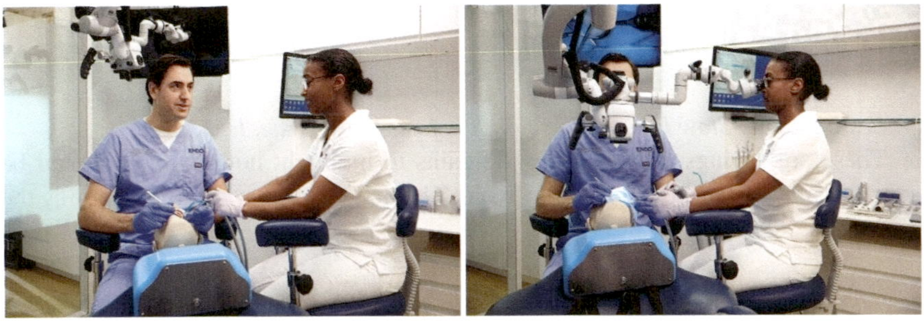

Fig. 4.7 Training drills without patients

concepts are the foundation of the procedure and need to be properly executed before procedural steps are rehearsed. By focusing on these fundamental habits and incrementally adding complexity, the skillset can be developed and built upon in a robust manner. The next steps are simple movements (e.g. single instrument passes) then complex movements (instrument exchanges), before rehearsing the entire procedure.

Every training step and instruction should be precise, direct, and communicated clearly, so that the task at hand is understood by all parties. When direction is vague, that leads to vague results. Imagine you had a farm with an empty field and you were training someone to plant seeds, in order for crops to grow and be harvested. Consider two sets of instructions, the first being, "go and plant these seeds in the field and make sure you use them all up". The second set is, "divide the seeds evenly, so that you plant them in rows 50 cm apart. In these rows, plant the seeds with a 30 cm gap between each crop, ensuring you have at least 2 seeds in each space". If you were the student, which set of instructions is easier to follow, clearly explains the intended outcome, and gives you a frame of reference so that you can evaluate your own performance? There is no doubt that the second set is the clearer. Being the teacher allows you to choose the type of instructions you give to your student, whether they be direct and precise, or vague.

The most honest method for students receiving feedback on their training and clinical procedures is with video recording. With varying angles, the footage shows exactly what movements occurred—in its purest form without bias. Modern-day smartphones and wide-angle action cameras (i.e. GoPro) give students the ability to quickly review clear videos of themselves training. This gives both the operator and assistant awareness of their actions that they never would have seen without it (Fig. 4.8). Reviewing footage as a team allows both members to analyse each other's movements, identify errors and have the immediate opportunity to improve on them. Progression can be visualised easily by comparing previous and current videos of identical procedures.

Video link to training video:

Festina lente. Slow is smooth, smooth is fast. More haste, less speed.

These are sayings used in many domains to highlight how training should be executed and what types of errors to avoid. The first is to concentrate on learning the

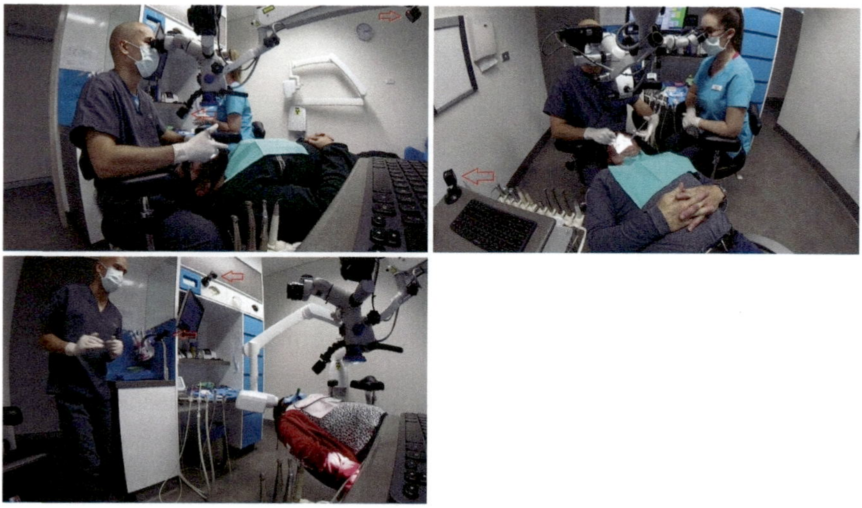

Fig. 4.8 Locations of operatory cameras for recording different angles

task slowly, which helps ensure you will be performing each step correctly. Getting it right is the most important, not being fast. If you focus on speed, it will lead to error for the inexperienced student, which will cost the team time. This is extra wasted time because the bad habit will need to be unlearned and the right skill subsequently learned. If you focus on rehearsing the procedure slowly, it will go smoothly. That will then speed the team's ability to execute the complex procedure. In the early stages of the training programme, it may seem as if there is little or slow progress. This should not be a deterrent. As long as each step is being done correctly, and mistakes are not repeated but learned from, that is the most important outcome. A mistake is valuable for the student because it means they now have knowledge of the right answer. You truly can only be wrong if you can compare it to what is right. When the correct procedures are executed correctly, only then will this lead to further proficiency and then speed.

References

Colvin G. Talent is overrated. London: Nicholas Brealet Publishing; 2016.

Dweck C. Mindset: the new psychology of success. Ballantine Books - Trade; 2008.

Ericsson K. An introduction to Cambridge handbook of expertise and expert performance: its development, organization, and content. In: Ericsson K, et al., editors. The Cambridge handbook of expertise and expert performance. Cambridge University Press; 2006. p. 3–19.

Ericsson KA, Harwell KW. Deliberate practice and proposed limits on the effects of practice on the acquisition of expert performance: why the original definition matters and recommendations for future research. Front Psychol. 2019;10

Ericsson KA, Pool R. Peak: secrets from the new science of expertise. New York: Houghton Mifflin Harcourt; 2016.

Gawande A. The checklist manifesto. London: Profile Books Ltd; 2010.

Horn J, Masunaga H. A merging theory of expertise and intelligence. In: Ericsson K, et al., editors. The Cambridge handbook of expertise and expert performance. Cambridge University Press; 2006. p. 587–611.

Training Programme

5

Abstract

The training programme should follow a structured sequence that begins with foundational lessons, which are further expanded upon as more skills are learned. This chapter will describe those in detail and provide specific steps and instructions on how to perform each procedural movement.

Key Points

- Training tasks should be rehearsed in sequence, starting with the simplest component of every movement, to the entire procedure.
- Situational awareness and pattern recognition are skills which form the basis of expertise.
- Limit motions to class 1 and 2 where possible.

The purpose of this programme is not only to prepare all team members for all endodontic-related tasks, but also how to recognise and adapt to situations outside of the main procedures. To do so, we follow the vision described in the previous chapter—which is to first train for every procedure without a patient. No clinical tasks are learned "on the fly"; they are all rehearsed before patient consultation and treatment. This is no different to when dental students undergo pre-clinical training in a simulation lab, before treating patients. When unprepared staff enter the clinical domain, errors are more likely to occur—some of them with very serious consequences.

Understanding why every step is performed in a particular way is the first critical stage in ensuring every team member has a clear expectation of their role in the team. Learning each step starts with a breakdown of every movement into its simplest components, which then cluster to become a single team skill. These are further developed to become an entire appointment simulation—a logical combination and sequence of all the skills already learned. Just like in the performing arts, a full

K. E. Yapp, C. Aznar Portoles, *Team Microscope Endodontics*,
https://doi.org/10.1007/978-3-031-70188-7_5

dress rehearsal is undertaken before the show starts. The reasons for training a complete procedure simulation are clear: (1) every team member knows exactly what to do, why the tasks occur, and how to perform them in the correctly anticipated order; (2) the patient can feel comfortable knowing the clinical staff are competent; (3) wasted movements are minimised or eliminated, thereby improving efficiency; and (4) everyone can have a happy, safe, and fulfilling clinical workplace environment.

A training programme should have a defined structure with clear learning goals for every training session. Whilst the teaching plan and methods of communication and implementation are important, the main focus should be on the outcome for the student—whether new skills are learned. Learning has been defined as "a relatively permanent change in behaviour potentiality which occurs as a result of reinforced practice" (Hilgard and Marquis 1961). The critical issue is the permanent change in the student in being able to perform the skill. *If it is not permanent, then learning has not occurred.* The task will then be able to be repeated without deviation from its description. To achieve this, deliberate practice (Horn and Masunaga 2006) is undertaken. This type of practice is focused, structured, and evaluated by identifying and eliminating errors. It involves immediate performance feedback and the opportunity to implement changes, with deliberate practice being carried out over extended periods of time.

Improvement occurs only when change can be implemented immediately and any performance errors are eliminated. Repetition is only valuable when the procedure has been mastered; if correction of each repeated task is required, repeating mistakes has no benefit and can work against the student by creating bad habits. The objective should be incremental daily improvement, where today's performance is better than yesterday's, with the aim of tomorrow's performance being better than today's.

Individual task performance analysis is important because it tells the teacher what the student is focusing on and how they adapt to the training situation. Understanding why such developments in training occur is arguably more important than the details of the performance task. The two broad areas of evaluation to be aware of are systemic issues and precision movements. Systemic issues relate to broader movements, general positioning, and operatory layout. Precision movements are related to fine placement of instruments and materials in and out of the operating field. The overall picture should be focused on first. Self-evaluation by the student should also be a skill to develop—without proper awareness of one's own performance and change over time, it will be difficult to improve efficiently.

Feedback should be focused at the task components and not the person. The very first skill taught to assistants—the mirror change—is effective for both parties to analyse. The issues to be highlighted are: listening and adaptation, broader and precision movements, and the skillset components receiving focus and attention.

Video link to mirror change video:

There should be two specific goals for both parties when giving and receiving feedback (1) developing situational awareness and (2) implementing immediate adaptation with correction.

Being aware of the situation *and* one's own performance are keys to developing expertise. Metacognition is a trait that experts have—they are aware of their own thought processes and the patterns behind them. Using the mirror change example, this will focus on specific aspects to analyse in the feedback process.

The mirror exchange does not aim to just provide the operator with a clean mirror. Neither is the aim to only ensure the new mirror is in the same position as the old mirror, prior to removal. The bigger picture goal for this and any task is to ensure that the team can continue the procedure, without delay. When there is a break—or potential break—in sequence, both team members should be aware of why it occurs, and have the ability to efficiently complete the task without wasted movement or time.

Before going through the four steps of the mirror change, everyone must begin with the end in mind: to seamlessly ensure that the procedure can continue by replacing the dirty mirror with a clean mirror, in the same working position for the operator. This is done without disruption to either team member. The steps are in the order of: (1) pinky tap, (2) pinky grab, (3) place new mirror, (4) release (Fig. 5.1).

Assistant preparation before this occurs should aim at: using the assistant binoculars to determine that the need for mirror exchange exists, identifying a suitable time when there is a pause in handpiece or ultrasonic use, and ensuring the new mirror is held in the correct pen grip position. These elements should be clearly communicated prior to rehearsal. Because all expectations of each team member are well defined, the feedback should focus on the preparation details. Some examples include reminding the assistant to hold the new clean mirror with a pen grip,

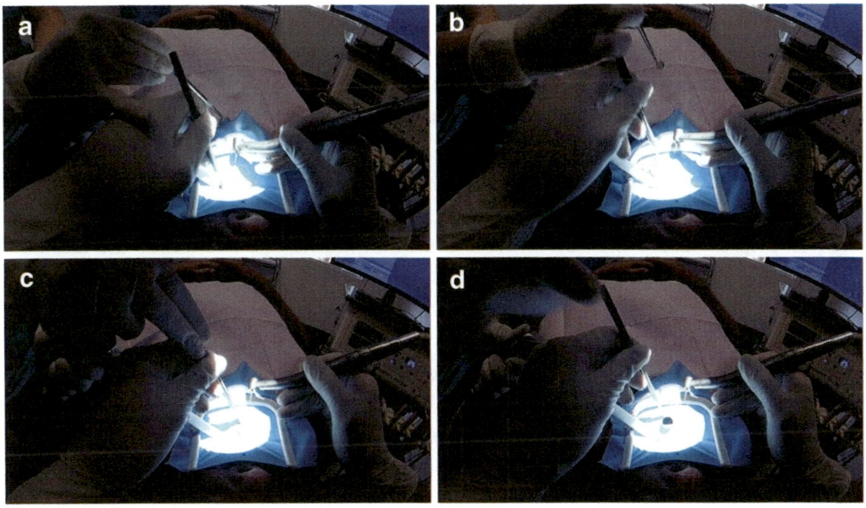

Fig. 5.1 Mirror change steps (**a**) pinky tap, (**b**) pinky grab, (**c**) place new mirror, (**d**) release

immediately after the suction is placed in the operating field. That way there will be no delay when there is a requirement for mirror exchange. Another would be to remind the assistant to use the binoculars and focus on the mirror—as well as the suction being close to the handpiece, preventing aerosol on the mirror—meaning the assistant can immediately determine the exact moment a mirror change is required and the right time to exchange. This proactive behaviour should be encouraged and is a sign of competency. If there is uncertainty about any part of the procedure, questions about specific scenarios should be asked and answered—i.e. change is required when water spray is covering the part of the mirror that visualises the bur, change only when there is a stop in handpiece usage, do not change when the handpiece is running.

The pinky tap is to communicate to the operator that the assistant is going to exchange mirrors. It is not to seek approval or permission. The purpose is to give notice to the operator to be ready to release the mirror and that this is just another step for the overall procedure to continue. If the old mirror is not released in response to the following step (pinky grab), this is due to improper communication. The feedback would be aimed at both assistant and operator. This would include: ensuring the pinky tap is firm enough to send a clear message to the operator, and for the operator to be aware and ready to release the mirror once the pinky tap occurs.

The pinky grab is a method of non-verbal communication. Once a firm grip has been secured by the assistant, that amount of pressure would be felt immediately by the operator, who then should respond by slightly opening and releasing the fingers to allow the dirty mirror to be removed. The old mirror should be removed towards the patient's feet and not towards the assistant's body, to avoid catching the mirror on the operator's index finger. Feedback for the assistant would be focused on the awareness of where and how the old mirror is being positioned once the operator releases the grip. More importantly, the dirty mirror should not be removed too far from the operating field, so that the new mirror can be replaced without wasting additional movement. For the operator, attention should be faced at ensuring the thumb, index, and middle fingers do not open too far, so that the assistant has a clear reference for where the new mirror is to be placed. Video recording and playback is useful in this situation, where precision and exact knowledge of correction is required.

Placing the new mirror should be in the same position with same orientation as the previously removed dirty mirror. The operator would then close the fingers into the pen grip to secure the new mirror. Common inefficiencies include the assistant placing the clean mirror with a different orientation, only to have to correct and change the position once placed. Feedback would aim at not just the assistant placing the mirror in the correct position, but to begin with holding the mirror with the right orientation in the first place. The operator would also need to ensure that once the mirror has been placed in the ideal position (the "lead"), the pen grip closes and the mirror is securely grasped (the "response").

The mirror release, by the assistant, only occurs in response to the firm grip of the new mirror by the operator. If there are non-verbal communication errors with the risk of dropping an instrument, either party should convey that to each other and then rehearsal with the intention on working on applying a firmer grip should occur. Doing so with the operator's eyes closed is critical in gaining a good understanding of how ideal touch perception should occur. Both assistant and operator should give each other feedback during the training process if the aim is to develop an expert team. The sequence for all tasks is: communicate clearly, rehearse, then provide feedback to reinforce and adapt to the originally communicated skills.

Once a skill is mastered by both assistant and operator, a role reversal is critical for each person to understand the other person's role which helps improve their own performance. This will instantly give a valuable perspective on positioning, firmness with exchanges and movements and timing sequence. The value is difficult to convey in words, however in our experience, every assistant and operator that has been trained with this understanding of the other person's role has shown immediate improvement and a greater understanding of the team skills—that were never initially understood.

The sequence of skills in the training programme is designed to begin with simple tasks that are easy to determine what areas the assistant is and is not focusing on. It quickly highlights what captures their attention and the aspects of the procedures that they value. You can immediately determine the "bigger picture" goals and performance indicators of a candidate early on in the training programme.

List of individual skills

1. Mirror change
2. 1–2 instrument pass
3. Irrigation
4. File passing
5. Local anaesthesia
6. Rubber dam
7. Capsule applicator
8. Composite applicator
9. Left hand skills
10. Cutting with scissors/scalpel
11. Suction handling and placement
12. Assistant Stropko use
13. Taking radiographs
14. Placing amalgam

Mirror Change

This is the first lesson taught to new assistants. It focuses on general and specific principles and because of that, is an effective way to begin assessment during a training programme.

One key message that is often overlooked is that microscope dentistry—for both operator and assistant—is a *completely* different skillset to working with an overhead chair light or loupes. Once this is accepted, then learning becomes easier and proper results are achieved in much less time.

With all teamwork exchanges, the underlying focus is that the movement should follow a lead and response. Clear communication occurs when there is a strong lead, which elicits a strong response. A response should not occur without a lead. If all members stick to the plan, not only will there be harmony with the movement, but there will be trust within the team. The ultimate aim of learning these specific skills is to develop trust between the operator and assistant, so that the procedure can be performed.

The aim of the mirror change is to allow the procedure to continue without disruption. The assistant exchanges a dirty mirror for a clean one without any difference in the working position of the mirror. That is, the new one is placed exactly as the old one was held. With all procedures, we begin with the end in mind. The assistant should be aware of the angle and position of the mirror, as well as where the operator is holding it along the handle. The operator holds the mirror in their left hand with a pen grip (Fig. 5.2) and only needs to slightly open and close the fingers to release.

Training breaks a procedure down into steps.

Step 1 is for the assistant to tap the hand of the operator with their pinky to signal that the mirror exchange is going to take place.
Step 2 is to grab the old mirror with the pinky and remove it.
Step 3 is to place the new mirror in the same position as the old one.
Step 4 is to release.

Because the assistant is looking through the assistant side binoculars, they will know when the mirror gets dirty. The assistant also knows when there is a quick break in the procedure. When changing a handpiece or pausing for inspection, this is the time to change mirrors. There should be no exchange whilst the handpiece is being used; otherwise, the operator will not be able to see.

Fig. 5.2 Left hand operator position as a signal to receive the mirror—the pen grip

Before the assistant starts the exchange, a few things should be noted. The new mirror is held higher up on the handle using a pen grip (Fig. 5.3). This is a neutral and relaxed grip and helps keep the shoulders at rest, as opposed to an overhand grip which tends to raise the shoulder and lift the elbow. The mirror is held in the position that the operator will use it. Where the mirror is facing, the handle position and angle—those details are kept the same as the old mirror. This is exactly what is done when passing a mirror at the start of a procedure.

Step 1 starts the team exchange and by tapping with the pinky, it indicates to the operator that they need to be ready and aware (Fig. 5.4).

Step 2 is the lead by the assistant. The mirror needs to be taken with a firm enough grip that the operator can feel that the assistant safely has control of it. This non-verbal communication is important. Because the operator is not looking out of the microscope or at their hands, the only way for them to be sure that the assistant has control of the old mirror is to feel it. Once the operator can feel that the assistant has a firm grip on the old mirror, they will let go, slightly, of the old mirror so the assistant can take it away. This is the operator's response to the lead. A slight release with minimal finger movement is important—because if the thumb and fingers were splayed and stretched out quite far, it would be difficult for the assistant to know where the new mirror should be placed. When the assistant has a firm grip on the old mirror, it is scooped up and brought out of the operator's hand in the direction of the patient's feet. This is because there is room to move here. If the old mirror is pulled directly towards the assistant, it will catch on and hit the operator's index finger. Pulling towards the assistant is an inefficient way to exchange that can potentially drop the mirror. There is much more space towards the patient's feet and a slight wrist rotation in this direction puts the hand in a more neutral, and slightly palm up position for the assistant. A slightly palm up position keeps the shoulder down and elbow lower. A more palm down position has the tendency to raise the elbow and shoulder. The operator is not holding the mirror high up on the handle. It is held lower, closer to the mirror head, to give room for the assistant. If the operator is used to holding the mirror higher up along the handle, this will prevent an efficient mirror change and cause a lot of wasted movement (Fig. 5.4).

Step 3 is the placement of the new mirror in the same position as the immediately removed dirty mirror. Note that this is the same as the initial mirror pass to the operator, to give the mirror to the operator in the first place, right after the patient is

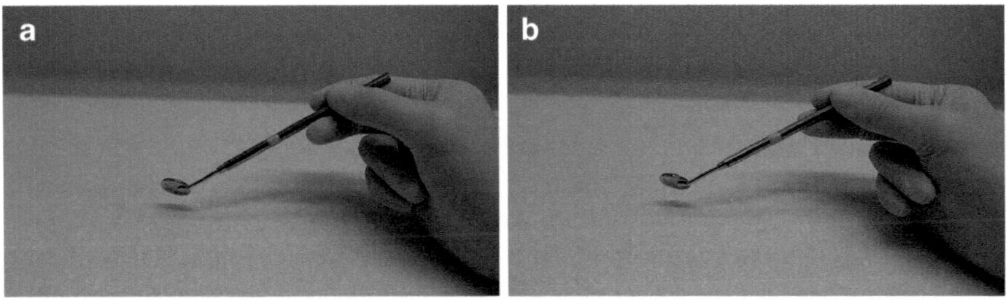

Fig. 5.3 Assistant holding the mirror using a (**a**) pen grip or (**b**) modified pen grip

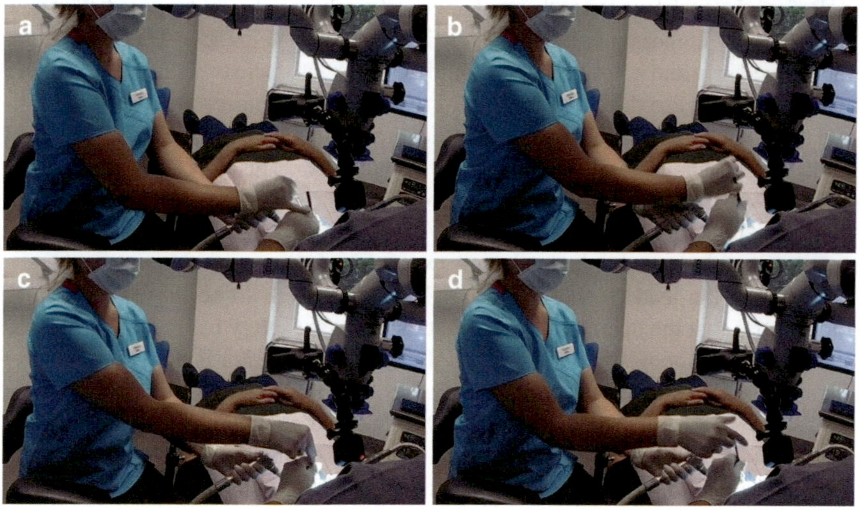

Fig. 5.4 Mirror change steps (**a**) pinky tap, (**b**) pinky grab, (**c**) place new mirror, (**d**) release

positioned lying down. The operator's left hand is in the pen grip position and the mirror handle is angled such that its position allows the mirror head to be facing the correct angle, positioned in the operating field, right where the operator requires it to continue the procedure. This is positioned between the operator's thumb and index finger. At the same time, the dirty mirror is being held with the pinky and secured against the palm in the same motion. The wrist should be in a neutral and relaxed grip, with new mirror placed firmly to communicate to the operator that the new mirror is in position (Fig. 5.4).

Step 4 releases the mirror in response to the assistant being able to feel the operator securing the pen grip. Once the clean mirror is in place, the procedure continues with the suction in the left hand placed immediately into the operating field, adjacent to the handpiece. During training, after placement, the assistant should first see if the new mirror was placed in the same position of the old one. This visualisation beforehand is exactly what the adage "begin with the end in mind" means.

Once an understanding of what changes are to be made (if any) occurs, skillset improvement and expertise development can occur (Fig. 5.4).

1–2 Instrument Pass

Almost all hand instruments are held and used in the same way: like a pen using a writing motion. This is the foundation for a safe and efficient instrument exchange. The method of the operator signalling to the assistant that an instrument is required is to hold their hand in the position that it will be used, otherwise known as the *signal to receive*. This is the non-verbal lead. When we talk about the lead and response, the lead by the operator applies to all movements and exchanges, which will then elicit a response from the assistant.

The pen grip, or modified pen grip, is the way that most hand instruments are held and used—with a pushing or scratching motion. This is also called the external precision grip because the instrument handle is external to the palm. The signal to receive a hand instrument is exactly as if you were about to start writing with an imaginary pen (Fig. 5.5). The hand is placed slightly outside of the operating field and close to the microscope light. This allows room for safe passing between assistant and operator. When the assistant notices the lead, they can respond by placing the instrument in the exact position that you would be using it, without any further adjustment. If that occurs, then that is a wasted movement.

There is a serrated grip on the handle indicating where to hold it and where to place it in the operator's hand (Fig. 5.6). The assistant should aim to place the instrument so that the operator's pen grip is holding it in the region of the serrated grip (Fig. 5.7). Too high and there is a lack of control. Too low and it cannot be secured nor used.

For a right-handed operator, instruments should be passed using the assistant's left hand.

At first, this may be a movement that a right-handed assistant is not used to, however with time and familiarity, development of a new skill will lead to a more natural and effortless movement. The reason for this is because passing with the left hand does not cross the field of view or risk hitting the operator's mirror or left hand, and the assistant does not have to reach or bend as far. Less wasted movement is better for every assistant and operator.

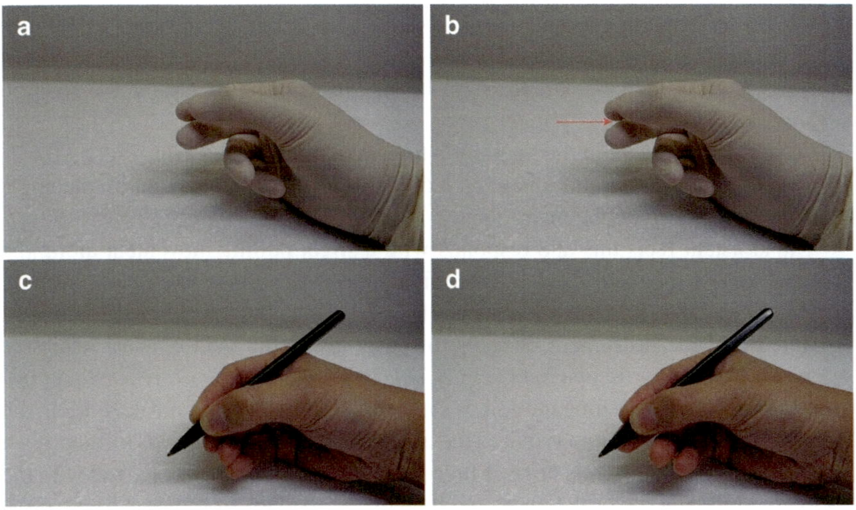

Fig. 5.5 (**a**) Pen grip—signal to receive, (**b**) arrow showing junction between index and middle fingers, where instrument should be passed, (**c**) pen grip, (**d**) modified pen grip

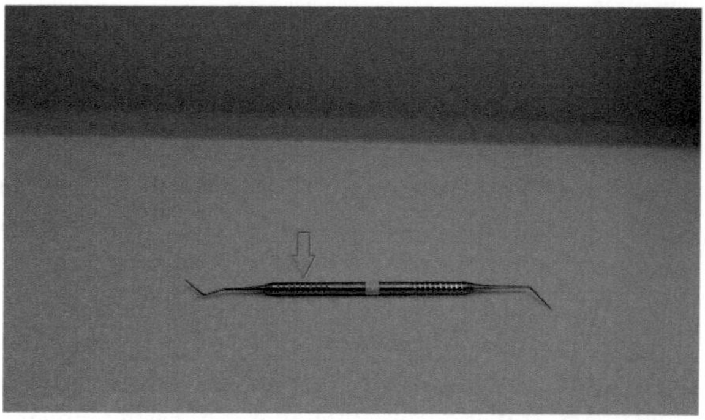

Fig. 5.6 Instrument with serrated grip (arrow) showing where to place it in the operator's hand

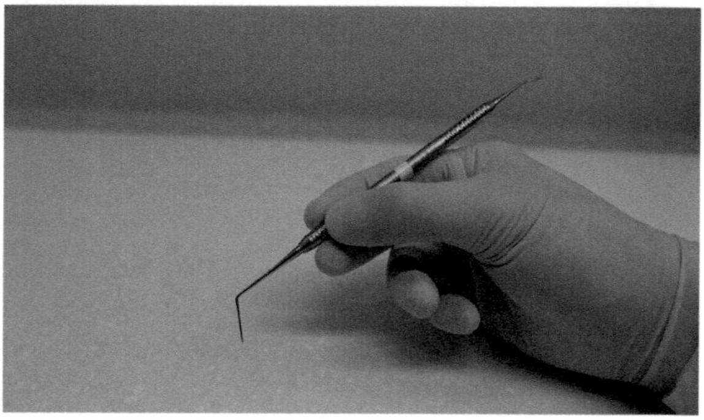

Fig. 5.7 Operator holding instrument on serrated grip area

For the assistant passing an instrument to the operator, these steps occur:

1. Use the left hand fingers and thumb to support the shank and top part of the blade of the instrument (Figs. 5.8 and 5.9).
2. Have a clear idea on where to place the instrument regarding its position and angle.
3. Place the instrument in the operator's fingers on the serrated grip (Fig. 5.10).
4. Once it is in the operator's hand, the operator responds to the lead by placing a firm enough grip on the instrument.
5. The assistant feels this and responds with a slight pulling movement along the long axis of the instrument into the light—an indication of the operator being able to visualise it in the microscope (Fig. 5.11).
6. Release of the instrument once the blade is inside the field of view, which is shown by the microscope light (Fig. 5.12). This is done for a few reasons. A pulling, rather than a pushing motion, is safer for the assistant to control. Holding the instrument at the far end requires the assistant to stretch and reach further. It's a small amount, but it adds up over time. The instrument cannot be passed with the blade in the middle of the operating field due to the lack of room and for safety. This whole motion reinforces the lead and response and eliminates confusion. The instrument is placed in the exact position and angle that the operator would use the instrument. This muscle memory will then become a habit, once learned and repeated, reinforcing safety and efficiency for the entire procedure.

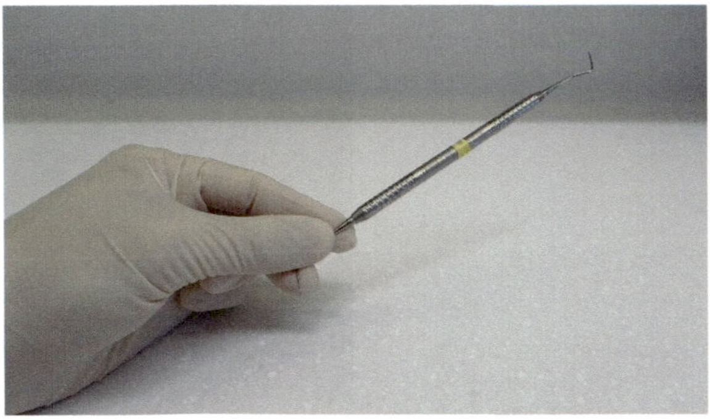

Fig. 5.8 Left middle and index fingers supporting instrument shank and blade with thumb on top

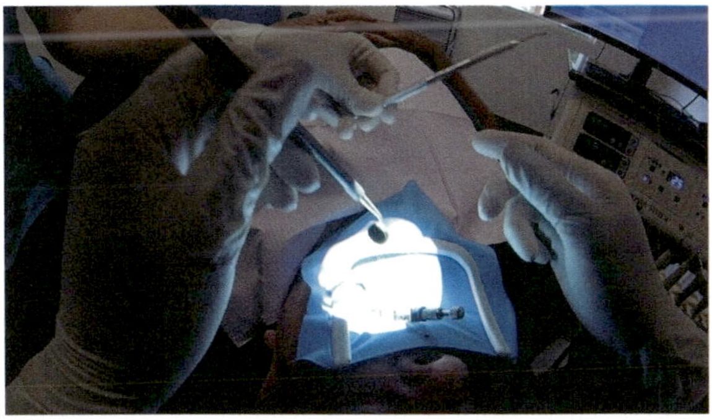

Fig. 5.9 Operator right hand in pen grip position as signal to receive

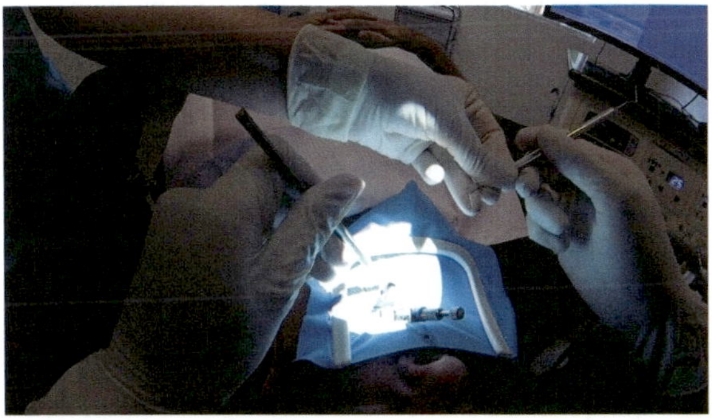

Fig. 5.10 Instrument placed into operator's fingers

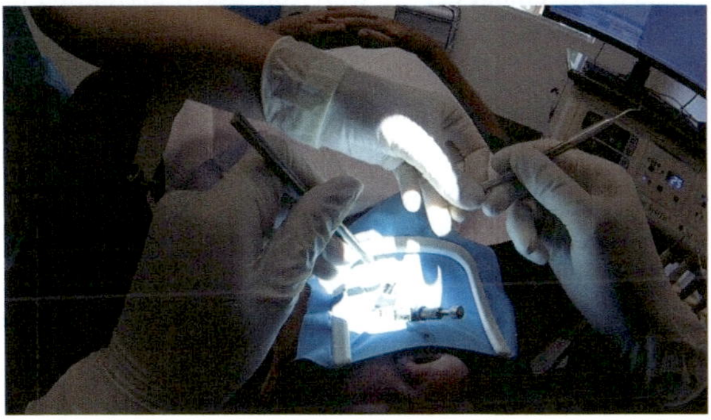

Fig. 5.11 After assistant feels operator has secure control of the instrument, it is pulled along the long axis with the blade into the microscope light

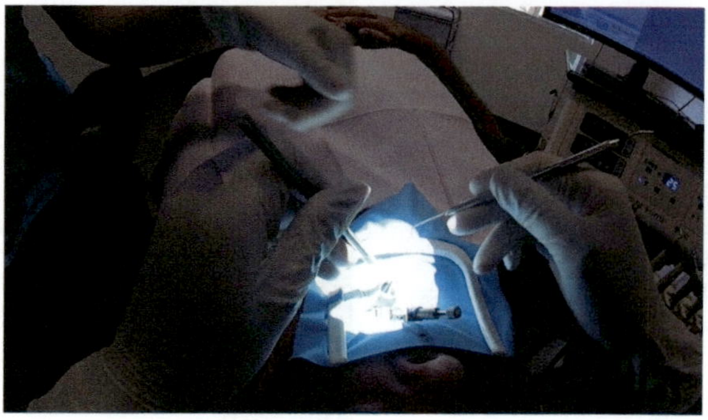

Fig. 5.12 Instrument released by assistant

Video link to instrument pass video:

When the operator passes the instrument back to the assistant, the "1–2" motion is used. There is no ambiguity with this. Every time the 1–2 is done, it means the operator is signalling to pass the instrument back to the assistant. If the operator does not do the 1–2, they are not passing the instrument back to the assistant and the assistant should not attempt to take the hand instrument from the operator. Traditional four-handed dentistry teaching used to emphasise the assistant proactively taking the instrument out of the operator's hand. This is a high risk way of thinking which can lead to injury and is not recommended because it is a dangerous movement. We advise that you never do this.

The 1–2 is as such:

"1" is a movement of the instrument upwards, mainly with the fingers and possibly slightly with the wrist.

"2" is a movement across, as if you were "giving the instrument back" to the assistant, mostly using the fingers.

When you, as the operator, feel the assistant has a firm grip, then you release. The firmer the assistant holds the instrument, the better. A controlled exchange is a safe exchange. The 1–2 is the lead, the assistant responds to the lead by taking the instrument securely, then the operator responds again by feel. If the assistant does not safely grasp the instrument, or the operator is not certain that the assistant has it securely and releases it, the instrument could be dropped. If this is a sharp instrument that falls on the patient, there could be disastrous consequences. Remember, the operator is not looking out of the microscope and at the hands—that is highly

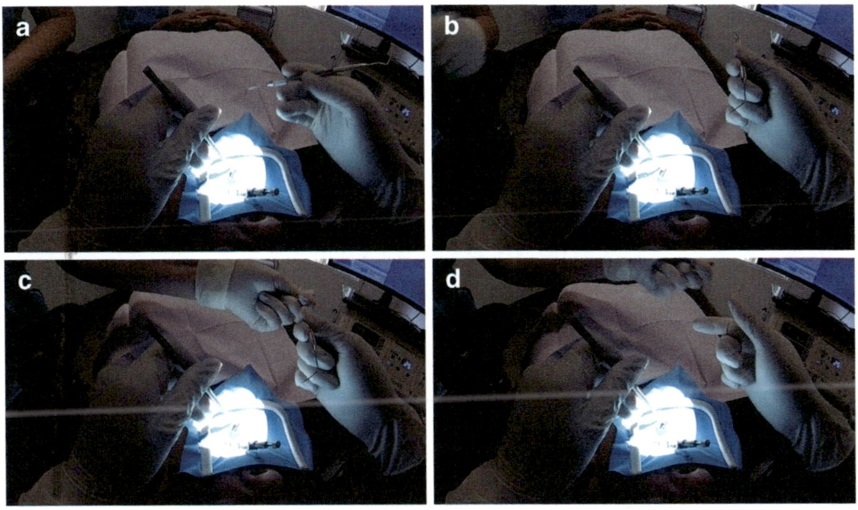

Fig. 5.13 1–2 Instrument pass (**a**) upwards movement "1", (**b**) forwards movement "2", (**c**) assistant grips instrument, (**d**) operator releases only after there is enough pressure applied by the assistant to indicate to the operator that there is enough control over the instrument

inefficient with a lot of wasted movement. It should be emphasised to the assistant that the operator not looking at the hands is an *advantage* to the team approach of safety and efficiency (Fig. 5.13).

Once the foundation of this movement is mastered by both operator and assistant, then the entire teamwork component to the procedure becomes easier. It is not so much the specific individual skill that is the potential issue, it is the understanding of how the team should and should not interact.

Locking tweezers follow the same movement, however due to the position of the clasp, the placement by the assistant in the operator's hand is slightly higher up (Fig. 5.14). The tweezers are also held by the assistant slightly above the blade, so the paper or gutta percha point is not being moved.

If the clasp is facing upwards, the tweezers are placed such that it only requires a slight movement of the thumb to unlock the tweezers. This position is for placing and removing gutta percha cones in lower teeth, as well as for paper points, wedges, cotton pellets, and other materials held by tweezers. The assistant should measure the cones by picking them up from a measuring block before passing. If the clasp is facing down, for upper teeth, the tweezers are placed such that it only requires a slight movement of the middle finger to unlock the tweezers.

If a hand instrument is being replaced with another one, this is done with one hand, in a single motion by the assistant. If not, the assistant has to waste time and energy by removing the old instrument, placing it down on the bench, taking the new instrument and passing it to the operator. First, the assistant has the new instrument ready in the position to pass. This is exactly the same motion as described above.

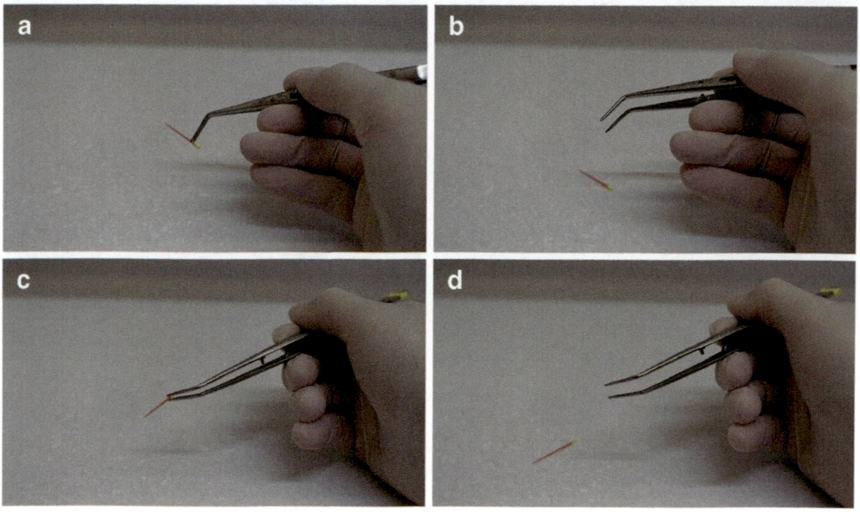

Fig. 5.14 Locking tweezers: for a mandibular tooth (**a**) passed so the thumb is adjacent to the clasp, (**b**) a slight thumb movement opens the beaks, releasing the gutta percha point, (**c**) for a maxillary tooth: passed so the middle finger is adjacent to the clasp, (**d**) a slight finger movement opens the beaks, releasing the gutta percha point

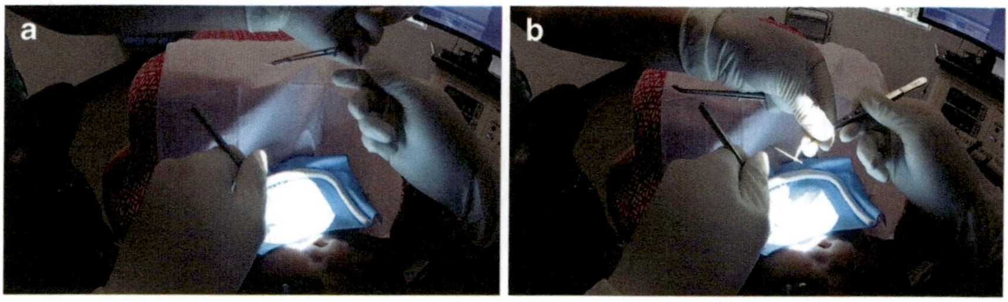

Fig. 5.15 One-handed instrument exchange (**a**) with the old instrument taken with the pinky finger, (**b**) the new instrument is passed as normal, with the old instrument being secured towards the underside of the palm

The 1–2 lead occurs. The assistant takes the old instrument between the pinky and ring finger of the left hand, brings it slightly towards the palm to secure its position, then passes the new instrument as normal (Fig. 5.15). With one motion, this is more efficient and prevents wasted movement. If the assistant takes the old instrument in their right hand, then places the new one with their left hand, that also works.

You will realise that many instruments are held and used like this—even for restorative procedures. Wedges in locking tweezers, fibre posts, tofflemire matrix retainers, micro brush for your bonding agent, bite block are all held with the pen grip and passed using the 1–2 motion. Other instruments like periodontal hand scalers and microsurgical scalpels also apply.

Irrigation

This method of using the index finger for irrigating and aspirating is not only more comfortable for the wrist, but also minimises wasted movements. What follows from a hand instrument exchange is the use of an irrigation syringe. Traditionally, dentists have been taught to depress the plunger with the thumb or the palm of the hand. This puts unnecessary strain on the wrist and cannot be used for all teeth. The neutral grip, with the index finger used to depress and aspirate the plunger is the most comfortable and efficient method that is used in all scenarios. This has all the elements of the 1–2 instrument pass with a slightly different grip by the operator (Fig. 5.16).

Syringe setup occurs with the needle being bent in line with the flanges, at a distance approximately one third from the hub, and at an approximately 45 degree angle (Fig. 5.17). With different mouth openings and different tooth positions, there

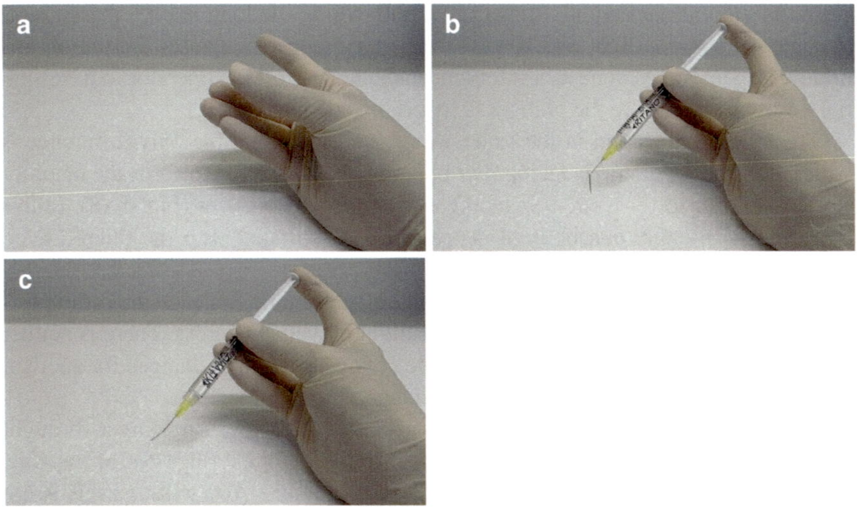

Fig. 5.16 (**a**) Neutral irrigation hand position. Irrigation syringe held for a (**b**) maxillary tooth, (**c**) mandibular tooth

Fig. 5.17 Syringe setup with the needle bent in the same axis of the flanges, one third from the hub

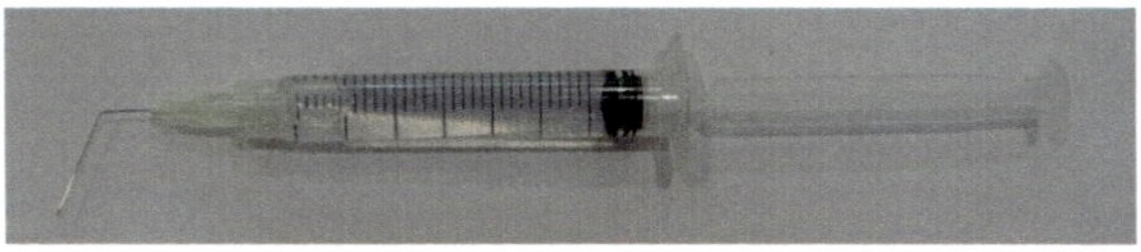

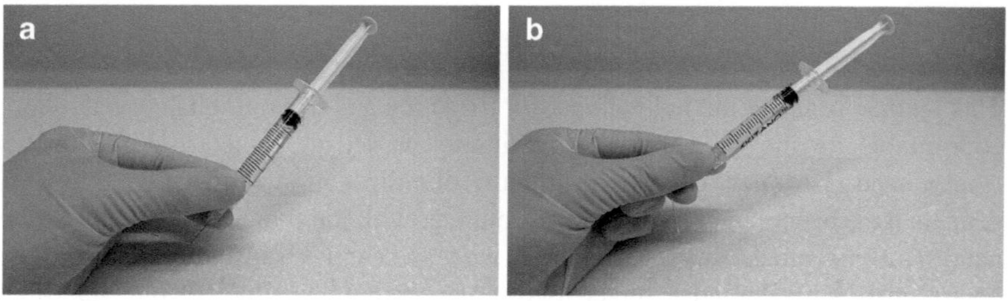

Fig. 5.18 Assistant holding syringe for (**a**) maxillary tooth, (**b**) mandibular tooth

may be some variation in the angle and height of the bend; however, this position is used for all teeth.

The signal to receive is with the thumb and middle finger at a distance slightly wider than the barrel of the syringe and positioned such that they will be able to rest against the flanges. The assistant holds the syringe at the hub, places it between the operator's thumb and middle finger, then pulls along the long axis of the syringe, into the microscope light and so that the flanges then rest against the thumb and middle finger. This is done as a safety mechanism, and for the operator to have spatial awareness of where the plunger is before it is depressed.

The syringe is passed in the same way for all teeth and the only difference is which way the needle tip is facing. For upper teeth, it is facing towards the operator (Fig. 5.18). For lower teeth, it is facing away from the operator (Fig. 5.18). In this position, the plunger can be both depressed and aspirated with the index finger, whilst keeping the syringe in the same position.

The 1–2 return is also done with the thumb and middle finger in the same position. "1" is a slight movement up. "2" is a slight movement across. There is neither additional nor wasted movement. When one syringe is being exchanged for another, it is very similar to a hand instrument exchange.

The old one is taken between the assistant's pinky and ring finger, brought slightly towards the palm to both secure and bring the old syringe out of the way, with the new syringe placed normally as described above. You will find this is not just for irrigation, but for etch, flowable composite, liquid dam, or any material in a syringe delivery tip. All of this is done with the operator looking in the microscope (Fig. 5.19).

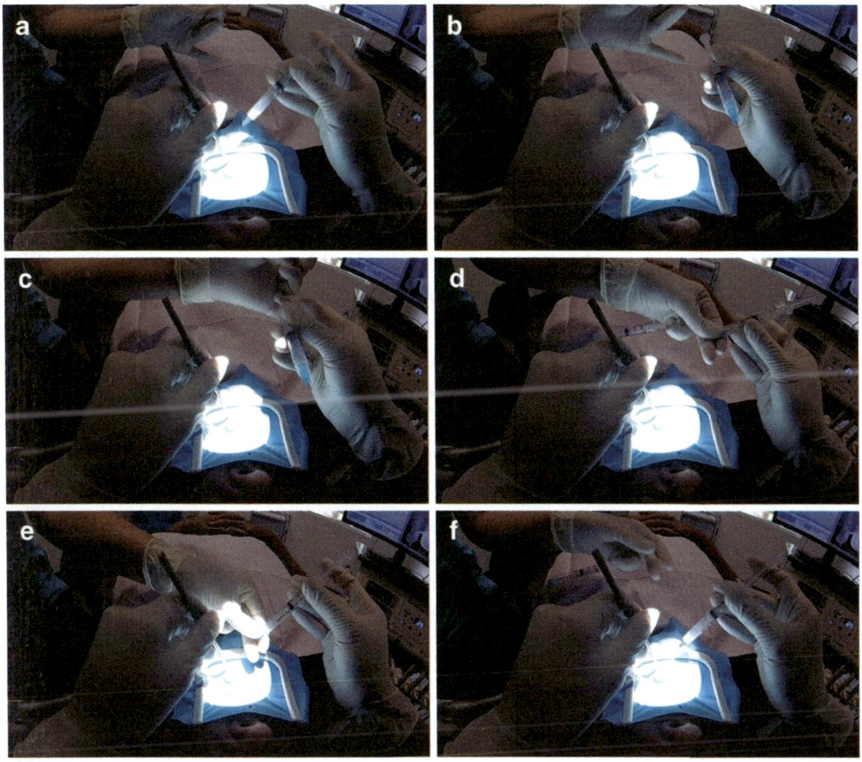

Fig. 5.19 Irrigant exchange (**a**) "1" motion by operator, (**b**) "2" motion by operator, (**c**) assistant securing syringe with pinky finger, (**d**) assistant placing syringe into operator's fingers with old syringe secured against the palm, (**e**) assistant pulling tip of needle into microscope light with flanges touching operator's thumb and middle finger, (**f**) assistant releasing syringe once enough pressure from the operator's fingers is felt, indicating sufficient control. Note this occurs with the assistant's right hand in the same position—with the suction tip still in the tooth

File Passing

The assistant controls the file movements, which is the safest and most efficient way to receive and return hand files. Many operators will have developed the habit of choosing and taking their own hand files. In a team microscope office, this is inefficient because of the additional wasted movement and lack of teamwork. Even if the operator has some accessory (ring) for them to hold the file sponge, this is a waste because it diverts attention away from the operating field by moving the mirror out of the way, taking the file, putting the mirror back and then using the file. Instead, the assistant holds the hand files in the sponge, then passes and receives the files by responding to the finger pressure from the operator. The hand files are not taken out of, nor put back into, the sponge by the operator. The assistant is placing the files into and taking them out of the operator's fingers by using the sponge. Once this mindset is understood, the movements become easier to learn.

The signal to receive is for the operator to hold the thumb and index finger together if they were to pinch the handle of a hand file—i.e. the way the hand file is held and used (Fig. 5.20). If you consider an imaginary long axis of the file, it should run from the operator's right hand to the assistant's left shoulder. This is just a guide, but as described earlier, if you have a specific goal directed task, it is easier to learn and correct.

The file holder sponge is held with the assistant's left index finger. The new file to be passed is placed on the side closest to the operator. This is so that there is room on the side furthest from the operator to place the old file being returned. The long axis of the file is pointed downward, from the assistant's left shoulder to the operator's right hand, so that the assistant has room to withdraw the sponge backwards and upwards. If the opposite occurs and the assistant tries to withdraw the sponge downwards towards the floor, there is not only insufficient room but also the risk of knocking the patient.

Once the assistant sees the thumb and index finger held together as the signal to receive, the hand file handle is placed between the thumb and finger, in the exact position that the file is going to be used. Only then, when the operator feels the handle is in the right position, a firm grip is placed securely on the file handle. When the assistant feels that the operator has control of the file, only then is the sponge withdrawn, with the file remaining in the operator's fingers.

The steps are:

1. Signal to receive
2. The file is placed in the operator's fingers
3. The operator grasps the file
4. The sponge is withdrawn

To return the file, the operator points it towards the assistant's left shoulder, with an identical position to receiving the file, then stays still. The assistant pierces the file into the sponge on the side furthest from the operator, who then releases the grip on the file after it is securely positioned and held by the sponge. The assistant then withdraws the sponge. If a new file is required immediately, that is already preloaded in the sponge and placed in the operator's fingers straight after the old one

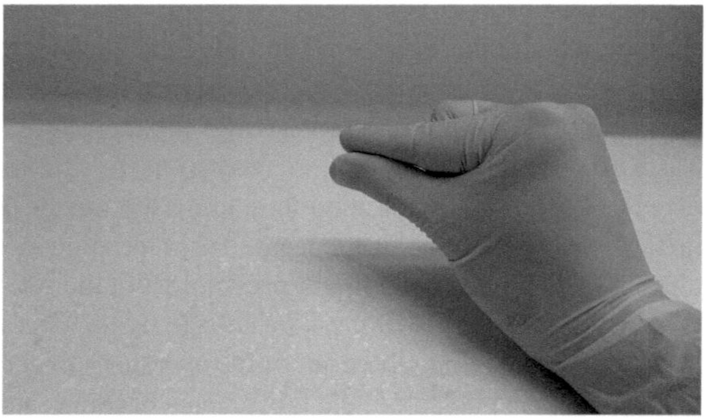

Fig. 5.20 Thumb and index finger held together as the signal to receive a hand file

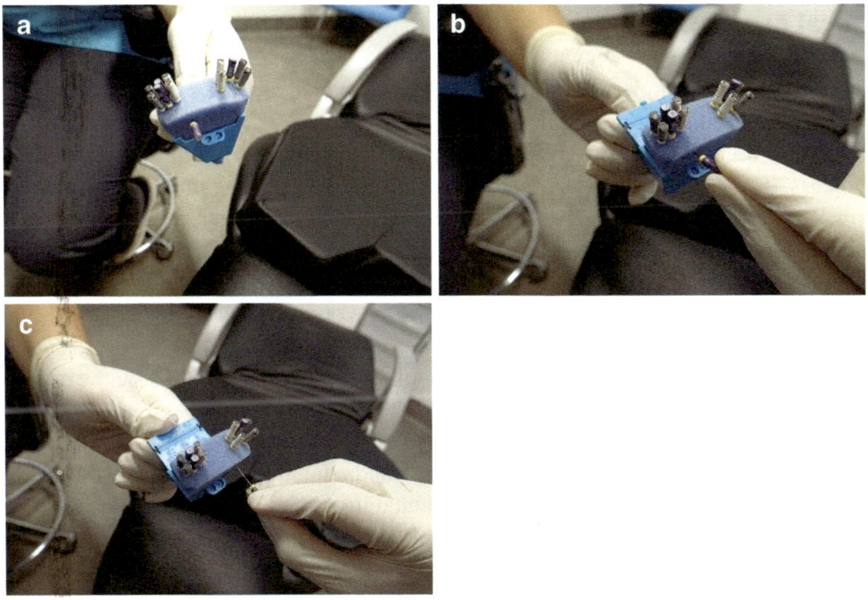

Fig. 5.21 (**a**) Assistant holding file ring with new file in position, (**b**) assistant passing file to operator, (**c**) operator pointing file towards assistant's left shoulder and returning file in side of sponge for old file

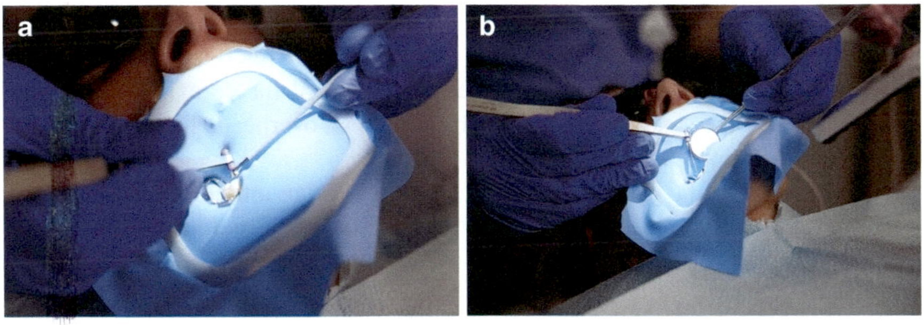

Fig. 5.22 File forceps (k1forceps.com) being used to instrument and control hand file

has been taken. All this is done in the same motion, so the assistant does not have to waste additional movements by removing the old and replacing the new file. This level of teamwork ensures that there is no disruption to the procedure and hand files can be exchanged quickly and safely (Fig. 5.21).

Another method of using and exchanging hand files is by using a file holder, i.e. the K1 file forceps (k1forceps.com). This is used like another hand instrument exchange in part (2) (Fig. 5.22).

Local Anaesthesia

Whilst not widely understood or accepted, team microscope anaesthesia allows for safe, relaxed, and precise administration of local anaesthesia. Learning why this method is used is the first step to successfully understanding and implementing the process. This is arguably the topic in microscope dentistry with the most controversy, the most debate and the greatest divide between clinicians. There is a lot of emotional protest and resistance against microscope anaesthesia, however once you overcome the mental fear and barrier of believing that you cannot administer local anaesthesia using the microscope, you will ask yourself why you did not do this earlier. You will ask why you refused to learn a new skill.

Just like everything else taught in this book, the main reason for using the microscope to give local anaesthesia is health and safety. Leaning and bending your back and neck is neither safe nor conducive to good health. Local anaesthesia requires precision, where support for your back, arms and shoulders is absolutely crucial in order to control the needle. Sulcular and palatal injections require delicate and controlled placement, which occurs when your body is well supported and at rest. Once you and your team learn how to give local anaesthesia injections with the microscope and refine your skillset, you have to ask yourself, what is the value in not using the microscope? Who benefits from taking away the microscope and using the naked eye to give local anaesthesia?

Some of the common objections that have not been shown to be logical nor true are:

- "I cannot see". If you have a proper low magnification microscope, you *can* see the whole mouth. You can also see better because of the co-axial lighting that does not give the shadows you would have with an overhead chair light. Having the wrong equipment is prohibitive.
- "I will miss and the patient will not get numb". For an untrained clinician, they can miss. Once you learn proper technique and delivery, there is no reason why a skilled operator cannot successfully give local anaesthesia with the microscope every time.
- "It's too slow". For those who have not learned how to use the microscope efficiently, all procedures will be slow. However once that skill has been developed, it is faster using the microscope because there is less wasted movement.
- "It will cause a needle stick injury". If all team members do not understand their role and how to exchange sharp instruments in general, then there will be a higher risk of needle stick injury. Once you train together and learn what to do, needle stick injuries will not occur.

Because this is a different sharp instrument, the team handles it differently. The key lesson for the team is to *only have one set of hands moving at a time*. This is a procedure where the operator uses both hands depending on the tooth. The left hand is for a lower right mandibular block and the right hand is for everything else—for a right-handed operator. A long needle is used for a mandibular block. A short

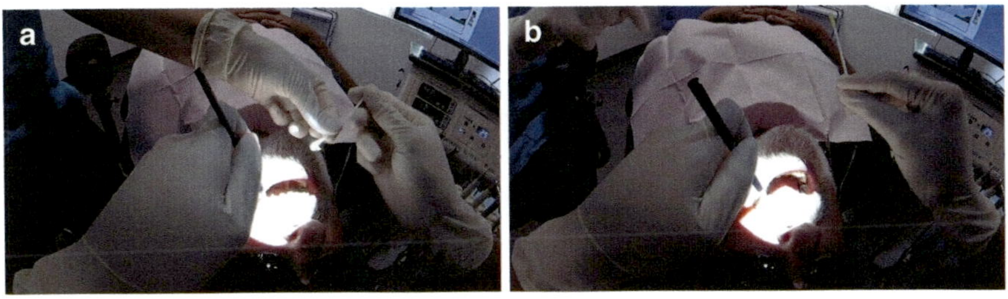

Fig. 5.23 Topical anaesthetic on a cotton bud being (**a**) passed like a hand instrument and (**b**) returned like a hand file

needle, bent in the same way as the irrigation syringe, is for everything else. The bend is in line with the flanges and one third the length from the hub, with a 45 degree angle. You hold the syringe the same way, but point the short needle towards the operator for maxillary teeth and towards the patient's feet for lower teeth.

We start with topical anaesthetic on a cotton bud. It is passed and used like a hand instrument with the pen grip and returned like a hand file—pointing towards the assistant's left shoulder (Fig. 5.23).

Right Hand Anaesthesia

The signal to receive is for the operator to have the palm facing up and with the hand still. The thumb, middle, and ring fingers are used to support the syringe. This is especially comfortable for those with smaller hands who may often have to overstretch their index and middle fingers. When the assistant can see that the operator's hand is still, then the syringe is passed. The aim is to point the needle towards the patient's opposite hip. This is an approximation, and as mentioned earlier, if you have a specific aim, it is easier to achieve the goal and make corrections. The body of the syringe is oriented parallel to the floor.

With the assistant's left hand, the palm is facing up and holding the syringe underneath close to where the operator's middle and ring fingers will be. The assistant then carefully passes the syringe to the operator so that their middle and ring fingers are securely around the syringe and the thumb may need to make a slight adjustment. The palm is still, the wrist and elbow are still, and the arm and shoulder area all still. Once in position, the operator will make a very slight squeezing motion with the middle and ring fingers—whilst keeping the palm still—so that the assistant can feel the slight pressure, as a form of non-verbal communication to indicate that the operator has safe control of the syringe. Then the assistant can release it and move out of the way completely, before the operator moves. All of this occurs with one set of hands moving at a time (Fig. 5.24).

The operator must have correct and relaxed positioning with the arms on the armrests. Make sure that the patient is in the right position and have them tilt their head back slightly if required, before receiving the syringe. The intended point of

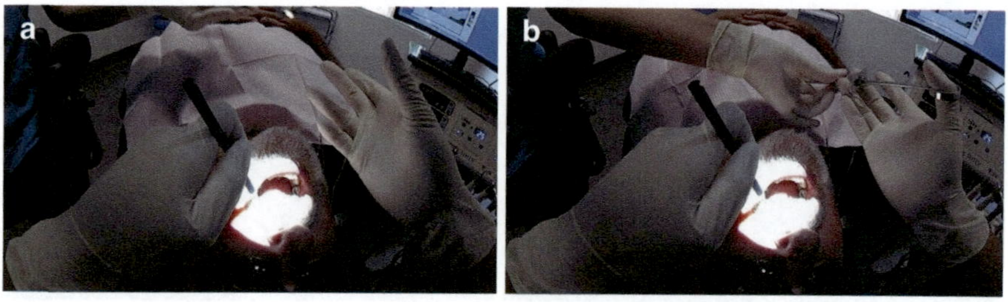

Fig. 5.24 (**a**) Signal to receive, (**b**) syringe being placed in operator's right hand

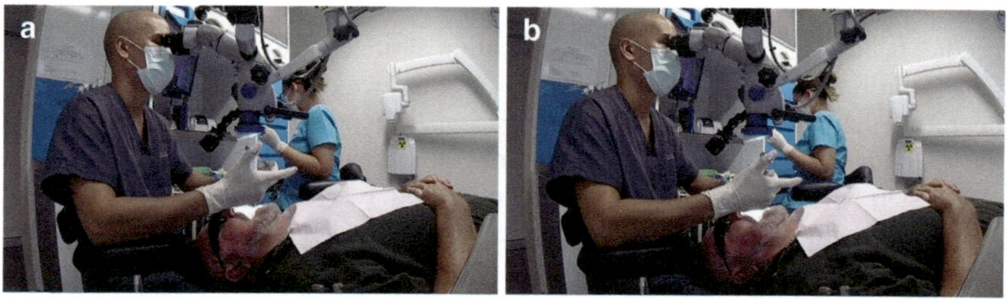

Fig. 5.25 (**a**) Operator receiving syringe and (**b**) moving the eyes only to look into the microscope, once the needle tip is into the field of view

needle entry should be close enough to the centre of the field of view—and not near the edge—to allow for any potential patient movement. This is why a reticle to aim on the eyepiece is valuable. The signal to receive is with the palm facing up and the hand still. The reason for having the eyes slightly back from the eyepieces is so that the operator only needs to move their eyes, and not the head, neck, back, and hips, to look out of the microscope and at the hands. The operator is looking at their hand just by moving their eyes only and can look back into the microscope with a quick eye movement.

The assistant passes the syringe as mentioned. After the assistant has released the syringe and their hand is well away from the operating field, the operator is still looking at the syringe before moving the needle tip into the microscope light. At this point, the needle is now in the operating field and can be visualised in the microscope. The operator immediately moves their eyes, looks into the microscope, and can clearly and safely administer the anaesthesia (Fig. 5.25).

For a maxillary infiltration, especially in the upper left or for a right palatal injection, the patient's head may need to be turned to the right, and the patient chair swivelled to the left, so that the arm can stay on the armrests and the patient's mouth remains in the field of view (Fig. 5.26). For a mandibular block, the angle of entry is generally from the opposite side premolars and in line with the maxillary occlusal plane—with the patient's mouth wide open. The mirror is used to retract, palpate the anterior border of the ramus, and also to apply pressure as a distraction mechanism to reduce the feeling of the needle puncture.

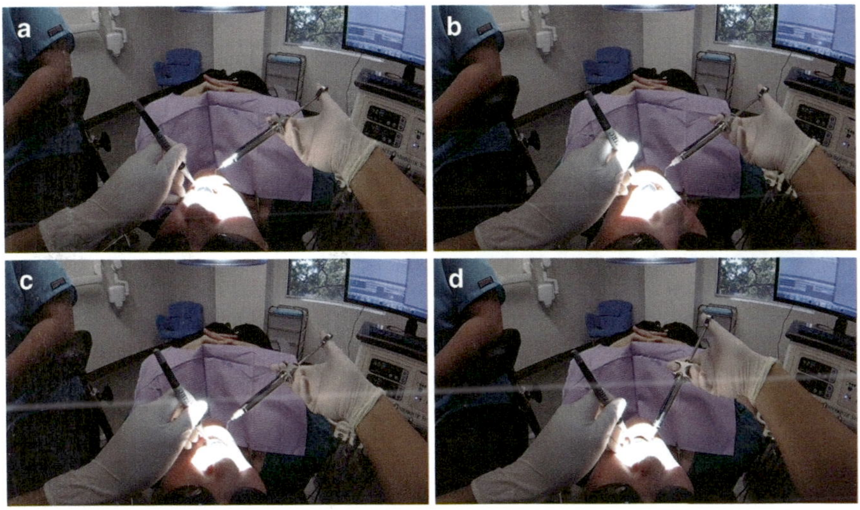

Fig. 5.26 (**a**) Patient in normal resting position, (**b**) head being turned to the right, (**c**) chair being swivelled to the left, (**d**) local anaesthetic being administered in a comfortable position with the arm on the armrest

Passing the syringe back to the assistant is the same motion as receiving, but in reverse.

Once the operator has finished administering the carpule, it is taken out of the mouth and held in the same position as when receiving. Palm up, needle pointing towards the patient's opposite hip, arm on the armrest and hand still (Fig. 5.27). When the assistant can see that the hand is still, their hand moves to take the syringe. Using the left hand, with the palm facing up, the syringe is held underneath, close to the operator's middle and ring fingers in the same position as passing the syringe. It is then taken away from the operating field. If a second carpule is used, that is done by keeping the left hand in the same position, taking the old carpule out with the right hand, putting the new one in, and repeating the process again, all without taking the left hand off the syringe (Fig. 5.27).

Left Hand Anaesthesia

The right mandibular block is done with the operator's left hand. This may be a foreign task for many, and the only way to get used to holding a syringe in your left hand is to practise. The more you do this, the quicker you develop the skill. The only difference for the assistant is the direction of the needle. In this case, it is towards the patient's right hip. The operator's left hand is still with the palm facing up and arm on the armrest. The syringe is held with the assistant's left hand underneath, palm facing up, but now with a slight rotation—pointing towards the patient's right hip. It is placed in the operator's left hand in the same way. There is a slight squeeze

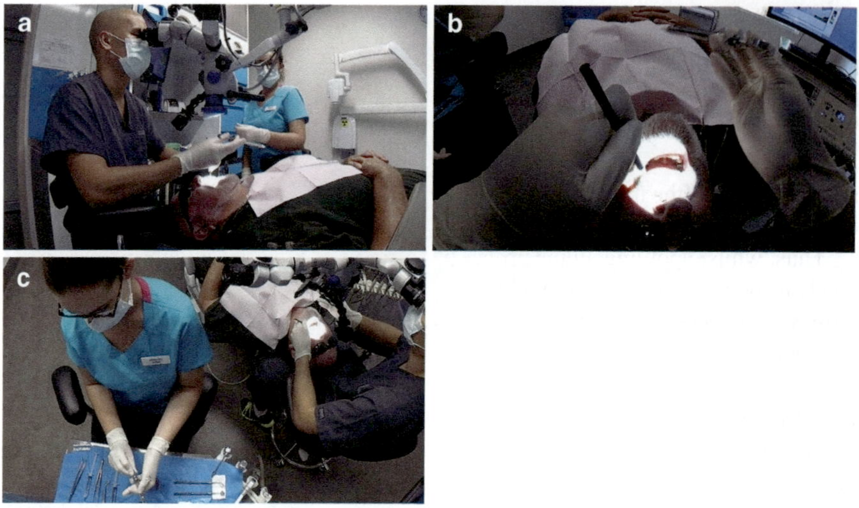

Fig. 5.27 (**a**) Returning syringe once empty has the same position as the signal to receive, (**b**) another view showing needle pointing towards patient's opposite side hip, (**c**) assistant changing carpule by keeping the left hand in the same position as receiving and passing the syringe

Fig. 5.28 Passing syringe into the operator's left hand

of the operator's middle and ring fingers. Once the operator safely has control of the syringe, the assistant releases and moves their hand away from the operating field. Then, the operator is free to move. This whole time the operator has been looking at the left hand, out of the microscope and only with an eye movement. The needle tip is moved into the light and then only the eyes move to look into the microscope (Fig. 5.28).

After administration, the syringe is taken out of the mouth and held in the same position as receiving it—pointing towards the patient's right hip (Fig. 5.29). The assistant does the reverse of passing, takes the syringe with the left hand with the palm up and close to the operator's middle and ring fingers. Changing the carpule is exactly the same—the left hand does not change position on the syringe and only the right hand is used to take the old one out and put the new one in. Passing the syringe again for the second time is the same as the first.

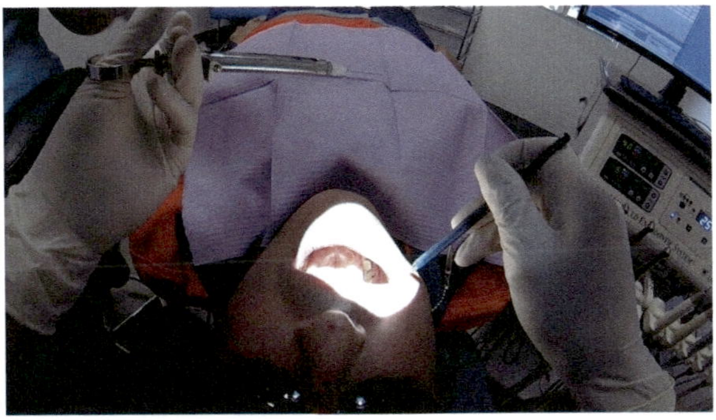

Fig. 5.29 Returning the syringe from the operator's left hand

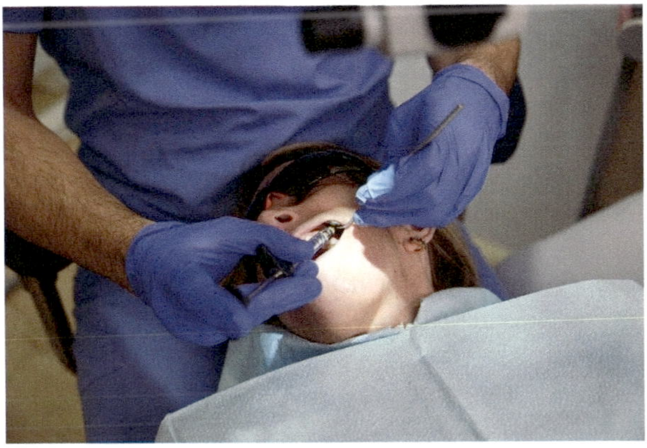

Fig. 5.30 Alternative local anaesthetic method

An alternative method is to hold the anaesthetic syringe similar to an irrigation syringe. That is to grip the barrel with your thumb and ring (fourth) finger, support the side of the plunger with your middle finger, then depress the plunger with your index finger (Fig. 5.30). The major limitation to this method is that if you have small hands, this will not work for you. Passing from the assistant to the operator is the same as the irrigation syringe. For a lower right mandibular block when the operator uses the left hand, the assistant holds the anaesthetic syringe with the left hand in the same way shown above. The assistant does not pass the anaesthetic syringe into the operator's left hand like an irrigation syringe.

You can now understand that the earlier objections may be true, only for someone who has not yet learned this skill. It is critical to understand and accept that microscope anaesthesia is completely different and not comparable to any other skill in dentistry.

Rubber Dam

The level of precision required for rubber dam placement is underappreciated, especially in partially broken down or oddly shaped teeth. Rubber dam placement with a team approach allows for precision and efficiency that would not be achieved without the microscope. The team approach involves efficient and precise movement with the least amount of wasted motion or unnecessary adjustment.

The frame, dam, and clamp are placed together in the one motion, with the mirror used so the operator can see the beaks of the clamp resting on the correct position on the tooth. A winged clamp is used in order to help stabilise the dam and guide it over the tooth and ensure that there is no catching under the beaks—which could lead to tearing during treatment. The angle of the forceps changes depending on the tooth, but the frame and dam is always passed in the same way—directly in line with, and exactly the way it is placed on the patient (Fig. 5.31).

Because the setup is done with all components as one, it is held with the assistant's left hand only. This allows the operator to hold the forceps in the right hand and stabilise the left side of the frame with the left hand (Fig. 5.32). There is no need for the assistant to use both hands and block the operator from holding and supporting the frame. The angle of the forceps is dependent on the tooth; however, all components can still be held with the left hand for all teeth.

The assistant passes the frame straight towards the operating field with the left hand only.

With a downward eye movement, the operator uses the left hand to hold the left side of the frame and the right hand to hold the forceps. The mirror is already held in the fingers, but shuffled down, in the left palm. Once the forceps are opened to stretch the clamp, it is placed over the tooth, the left side of the frame can be released and the mirror brought into the operating field, with a quick eye movement to precisely guide the clamp placement using the microscope (Fig. 5.33). When the clamp is placed in the right position and beaks released onto the tooth, the forceps are returned in a palm up position. The way the forceps face depends on the tooth, but the important message here is that the assistant should pay attention to this particular position—the signal to return the forceps. It is exactly the same position that will be used when the assistant passes the forceps at the end of

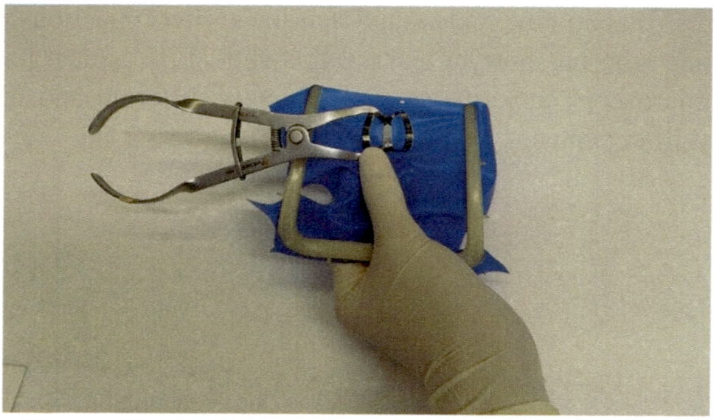

Fig. 5.31 Rubber dam, frame, and clamp held together with left hand

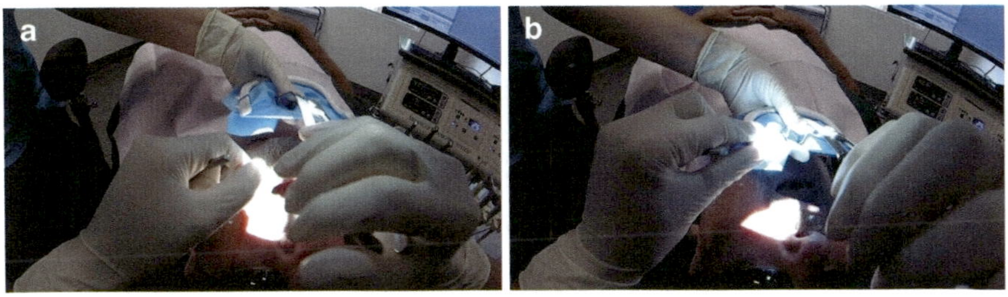

Fig. 5.32 Quadrant 2 molar tooth (**a**) assistant passing rubber dam, frame, and clamp, with left side of frame free for operator to hold, (**b**) operator's left hand securely holding left side of frame

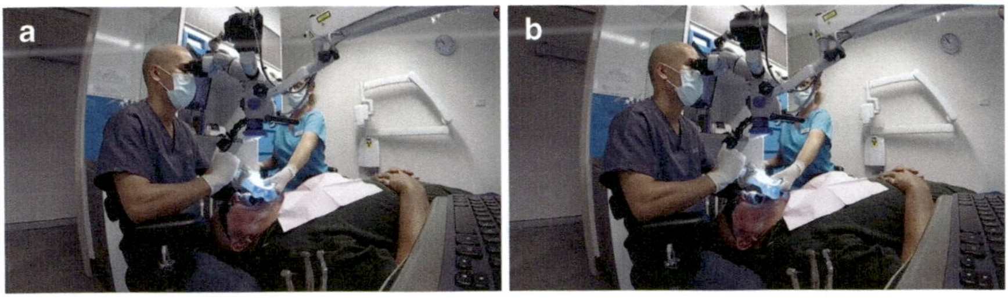

Fig. 5.33 (**a**) Operator looking at hands to take rubber dam forceps, (**b**) eye movement upwards to look into the microscope

the procedure to take the rubber dam off. It is the same movement but in reverse (Fig. 5.34). You might be noticing a trend here—the simplest way is to learn a procedure one way, not using multiple methods for the same procedure (Fig. 5.35). That only adds confusion.

Before the dam is stretched under the wings, the clamp is rocked with a firm movement to check the stability. If multiple teeth are being isolated, they can then be stretched and flossed as required. Note that the rubber dam is stretched tight across the corners of the frame on both the top, and bottom edges, leaving the space between the top and bottom edges loose and able to be stretched. This makes placement easier and once the clamp is in position, the dam can be quickly stretched across the corners of the frame.

Another factor to note is that the hole punched for the tooth is placed in a position that leaves the top edge of the dam positioned just under the nose and not irritating the patient. Importantly, this prevents any trimming of the dam around the nose or additional wasted movement. You may need to experiment with the position and use a stamp or make your own template. You also want to ensure that the top edge of the dam is not too far away from the nose such that there is a noticeable gap between the top of the rubber dam and the oral cavity, which could risk instruments or materials being swallowed. Limiting wasted movements is the key to efficiency.

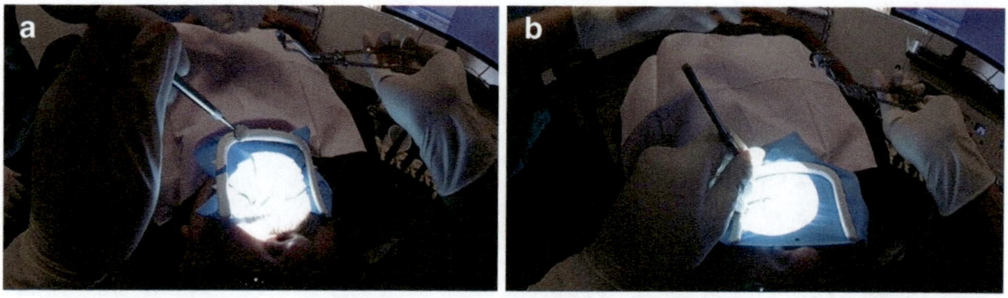

Fig. 5.34 Returning forceps immediately after placement with a palm up position. This is the same position for the assistant to pass the forceps at the end of the procedure

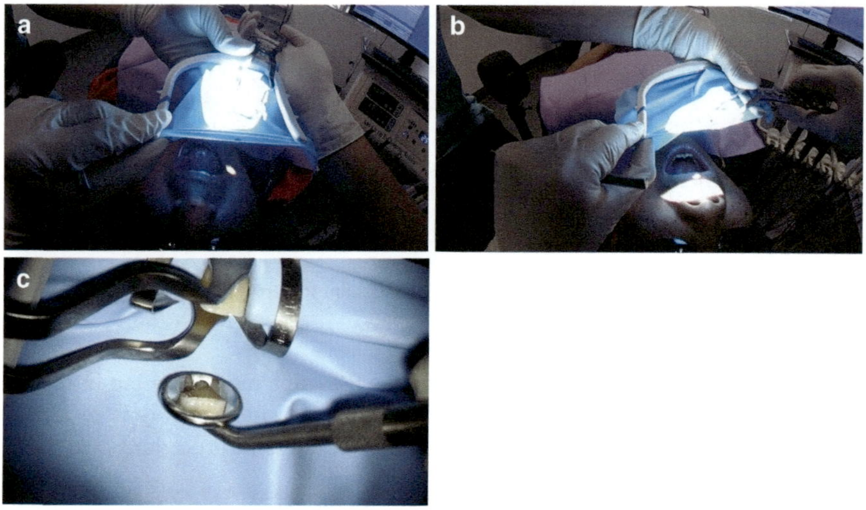

Fig. 5.35 (**a**) Passing position for quadrant 4 molar tooth, (**b**) passing position for quadrant 1 anterior tooth, (**c**) microscope view with mirror checking position of rubber dam clamp beaks

Capsule Applicator

The capsule applicator is mainly used for administering glass ionomer cement, often used as the occlusal layer of a temporary filling above the Cavit or other calcium sulphate-based temporary filling material. The applicator is held and used one way, which is with the palm open and facing upwards, with the button facing the operator. This is so the arm is resting on the armrest and the moveable activating trigger is squeezed with the fingers. This is also the signal to receive. The only difference in capsule positioning is to face the tip towards the button for an upper tooth, and away from the button for a lower tooth. Once the material has been used, a slight shuffle of the applicator up into the palm occurs, before a 1–2 instrument pass. This allows the assistant to have room to take the applicator by holding the handle and immediately pass the next instrument in one motion with the left hand. Often, a slightly damp cotton bud is used. If you wanted to use a microbrush, flat

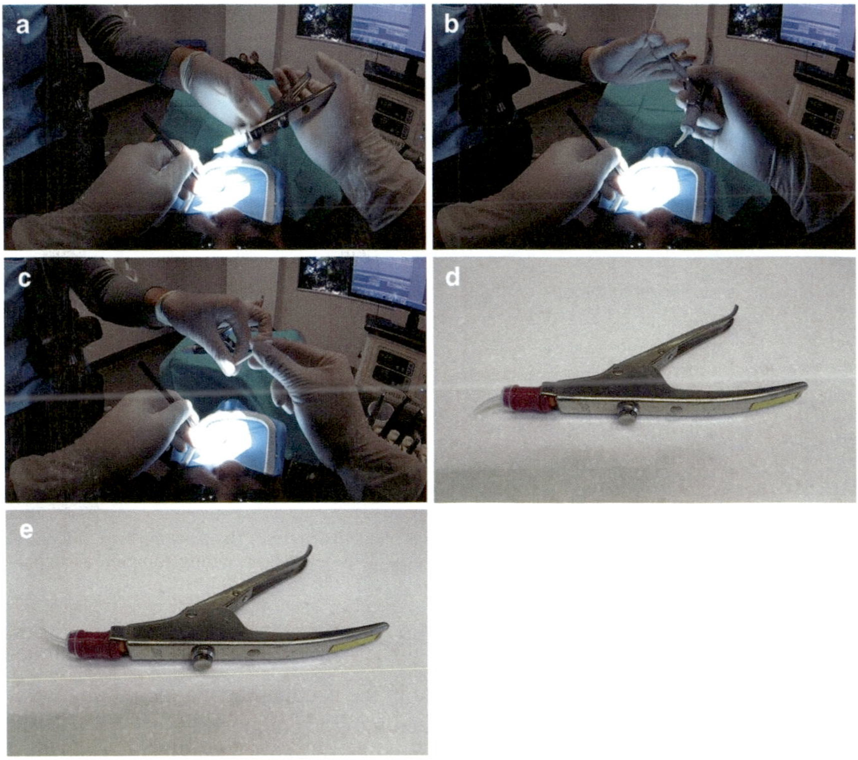

Fig. 5.36 (**a**) Passing the capsule applicator to operator, (**b**) returning to assistant, (**c**) assistant exchanging with a cotton bud in one motion, (**d**) capsule position for a maxillary tooth, (**e**) capsule position for a mandibular tooth

plastic instrument, ball burnisher, or other hand instrument, these are all exchanged, held, and used in the same way, using the same motion (Fig. 5.36).

Composite Applicator

Composite capsule delivery is different from that using glass ionomer cement because of the different applicator design. The main format is a needle delivery tip for cementation of fibre posts and core build ups using a dual cure composite. The purpose of the procedure is to fill the post space and core space with composite—with a fibre post immediately placed—that adapts to the margins without any visible voids. Because the depth of the post space can vary, it is important to check the length and angle of the needle tube prior to composite placement. The operator bends the tip and checks placement using the microscope. The measuring ruler allows for this to be done effortlessly.

Centrix AccuDose tips are the ones being referred to in this section; however, a similar sized alternative with identical dimensions may be used. The 20 gauge orange needle tube has a wider diameter and is shorter than the 24 gauge

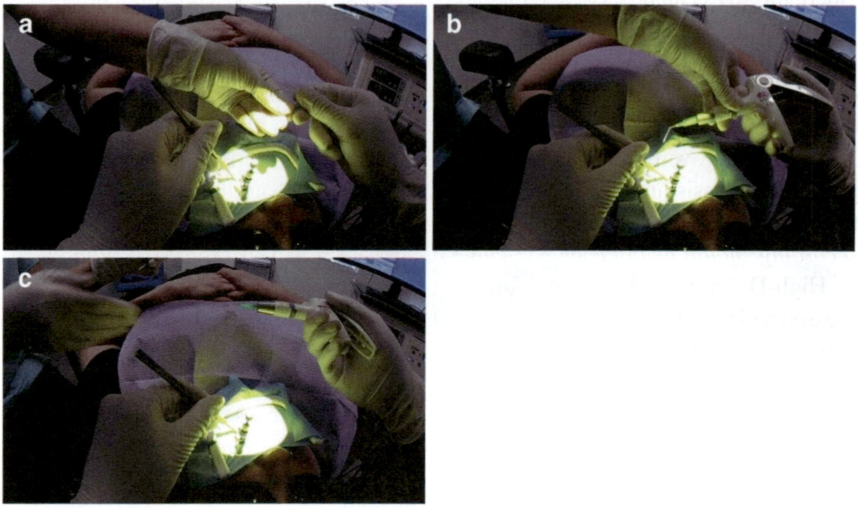

Fig. 5.37 (**a**) Passing AccuDose tip identical to a hand instrument, (**b**) passing loaded Centrix syringe, (**c**) returning syringe in the same way as passing

aquamarine tip. In many cases, the orange tip is neither long nor narrow enough and the aquamarine tip would need to be used.

When the assistant places the composite in the needle tube via the mixing tip, it is important to prevent air being incorporated into the composite. One way is to fill the needle tube from halfway up, thereby leaving space in the tip. The plug sits flush with the composite before delivery and the air gets pushed out first.

The signal to receive is for the operator to hold the palm up and open, similar to receiving the capsule applicator. The direction of the needle tip bend corresponds to the arch—either pointing up or down. The assistant places the syringe in the operator's open palm and moves it into the light, where the operator is continually looking in the microscope (Fig. 5.37). The composite is placed into the post space first, then the rest of the cavity before the operator passes it back to the assistant using the same open palm position. The pre-measured fibre post in the locking tweezers is then passed immediately—just like a normal hand instrument pass. The fibre post is placed to the correct length, before a dry microbrush is used to make any adaptation as required. Depending on the manufacturer's instructions, sometime chemical cure may be required before light curing. This step is done after the relevant etch, prime, and bonding steps have been completed.

Left Hand Skills

Right-handed operators use the left hand for tasks besides holding the mirror. Using the left hand not only gives the right hand some time to relax and is an ergonomic advantage in some situations. In particular, using the electronic apex locator, making bends with the tweezers and left hand filing.

Many people use the shepherd's hook attachment clip onto the hand file for the electronic apex locator. The disadvantage of this is that the rubber stopper cannot come close to the handle if clipped in that region, the common place for attachment. It also does not allow for easy simultaneous movement of the stopper and the file in order to obtain an efficient reading. An alternative to the bulky clip is the two pronged fork attachment, which comes as a long handle. Because it is thin, it allows for close placement of the rubber stopper to the handle, where you can move the file and the stopper together whilst quickly getting a reading from the unit. This is used as a pen grip in the left hand and allows the operator to control the file with the right thumb and index finger (Fig. 5.38).

Most importantly, the operator is focused and looking in the microscope and not at the apex locator monitor or device. The assistant is able to read out what's on the display by saying the number out loud. Erratic readings can occur if there is fluid or tissue in the canal or the file is touching a metal crown. The usual sequence is to establish a patent glide path, then dry the canal using needle aspiration and use a size ten file with the apex locator. Patient positioning is important because having direct vision to the occlusal surface of the tooth ensures that you can visualise where the file is positioned in relation to all the reference points on the tooth. If there is a metal crown, you can see if there is metal on metal contact that could affect your reading and make an immediate correction. All of this is so that you can work efficiently and avoid having to stop, get a mirror, use it, pass the mirror back, make wasted movements, and disrupt your focus.

The signal to receive is a left-handed pen grip. The apex locator fork is just a hand instrument. Before this, the mirror is slid down in the left hand and held with the ring finger and pinky. The right hand should have already placed the file in the

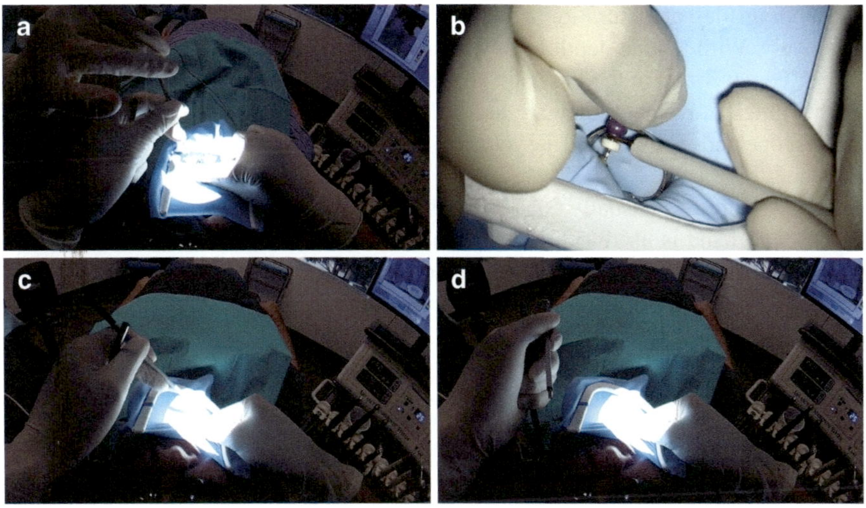

Fig. 5.38 (**a**) Assistant passing the apex locator fork attachment similar to a mirror, (**b**) left hand holding fork attachment whilst right hand is moving file, (**c** and **d**) returning using a "1–2" movement

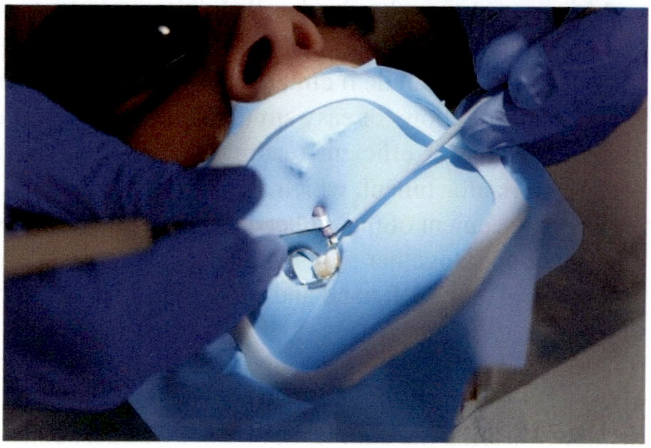

Fig. 5.39 Using apex locator with file forceps (k1forceps.com)

canal just short of the estimated length, and the fork in the left hand touches the file between the handle and rubber stopper. If it's short of the apex reading, the length is extended with the fork still touching the file. The fork can easily and quickly adjust the stopper when the reading is at apex, or whichever value you choose to aim for. If the reading is past the apex, doing the opposite (withdrawing the file) is done using the same motions. All of this is done with the operator looking in the microscope (Fig. 5.39).

Some apex locator units might be too small or the cord too short to be securely left on the bench. In this case, it can be held in the assistant's hand in a position where the operator only needs to move their eyes to look at the unit if required (Fig. 5.40). Often these units have recognisable sounds that indicate each position—meaning neither assistant nor operator needs to look at the unit.

Bending files or obturation needles with the tweezers in the left hand is done efficiently. With a wide depth of field, it is easily done without moving the microscope. The tweezers in the left hand are passed and held just like a hand instrument. With a quick movement, they are used to bend either a file or needle tip, all in the microscope and the tweezers are returned with a 1–2 hand instrument pass with the left hand (Fig. 5.41).

Left hand filing is usually done in lower left molars, after the file has been placed and the left hand is used to file using small hand files to obtain a glide path. This allows the right hand to rest; the left hand is in a more amenable position to hand filing with the palm faced slightly downward.

Cutting with Scissors/Scalpel

Operators will use scissors to make intra-operative modifications like cutting gutta percha cones during treatment. Other tasks like cutting matrix bands should be done during preparation and setup, prior to the patient entering the room.

Adapting and making intra-operative changes by cutting with the scissors is an important step in order to make precise adjustments to the gutta percha cone. In the

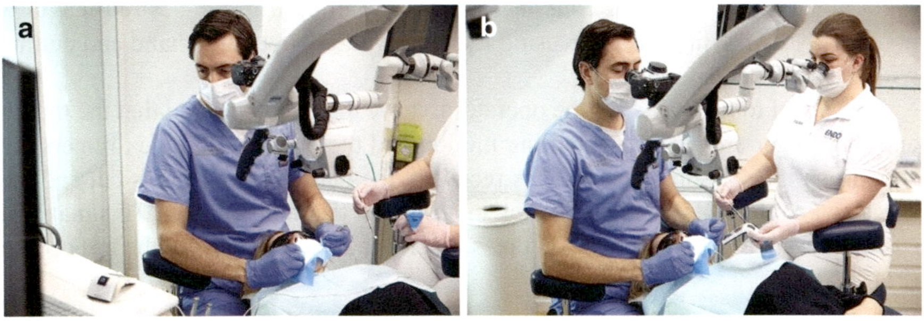

Fig. 5.40 (**a**) Apex locator unit should not be on the cart which requires a neck movement and class 5 motion, (**b**) instead the assistant holds it closely, so an eye movement is needed to look at the screen

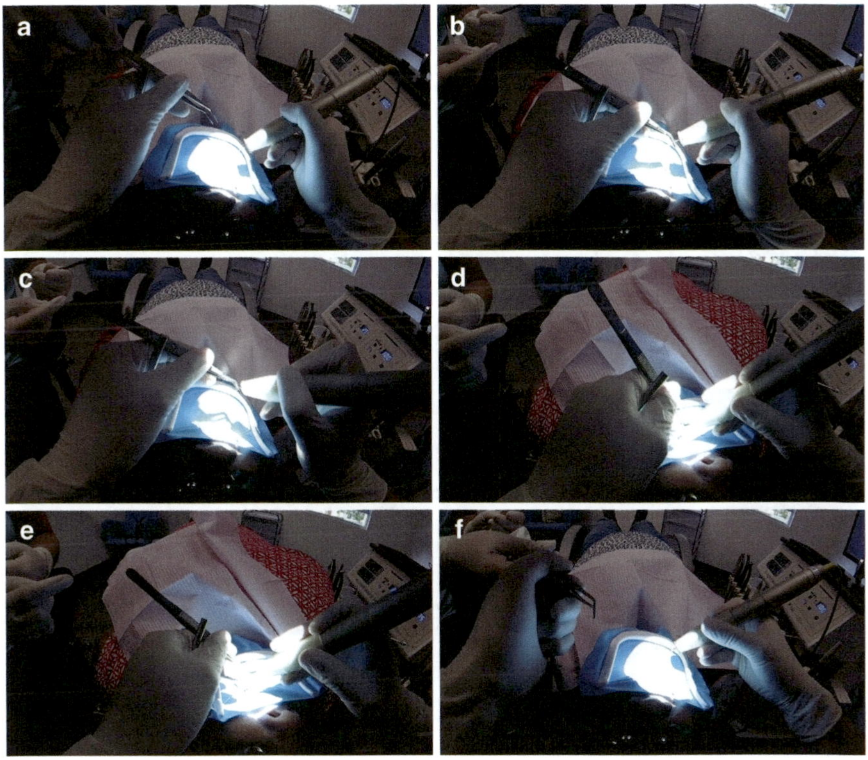

Fig. 5.41 (**a**) Tweezers passed into the operator's left hand similar to a mirror, (**b** and **c**) bending the obturation needle by moving the right hand, (**d** and **e**) bending the obturation needle by moving the left hand, (**f**) returning with a "1–2" movement

event you are long with the cone measurement, cutting it to the desired length is done easily by holding the cone using locking tweezers in the left hand and cutting with the right hand.

The signal to receive is for the operator to have the palm open, so that the thumb and ring finger are used for a cutting motion and the middle and index fingers are supporting the scissors. When this is received, the gutta percha cone is cut and the scissors are returned with a slight flick motion with the middle finger. This makes it easier for your assistant to take the scissors and continue with the procedure (Fig. 5.42).

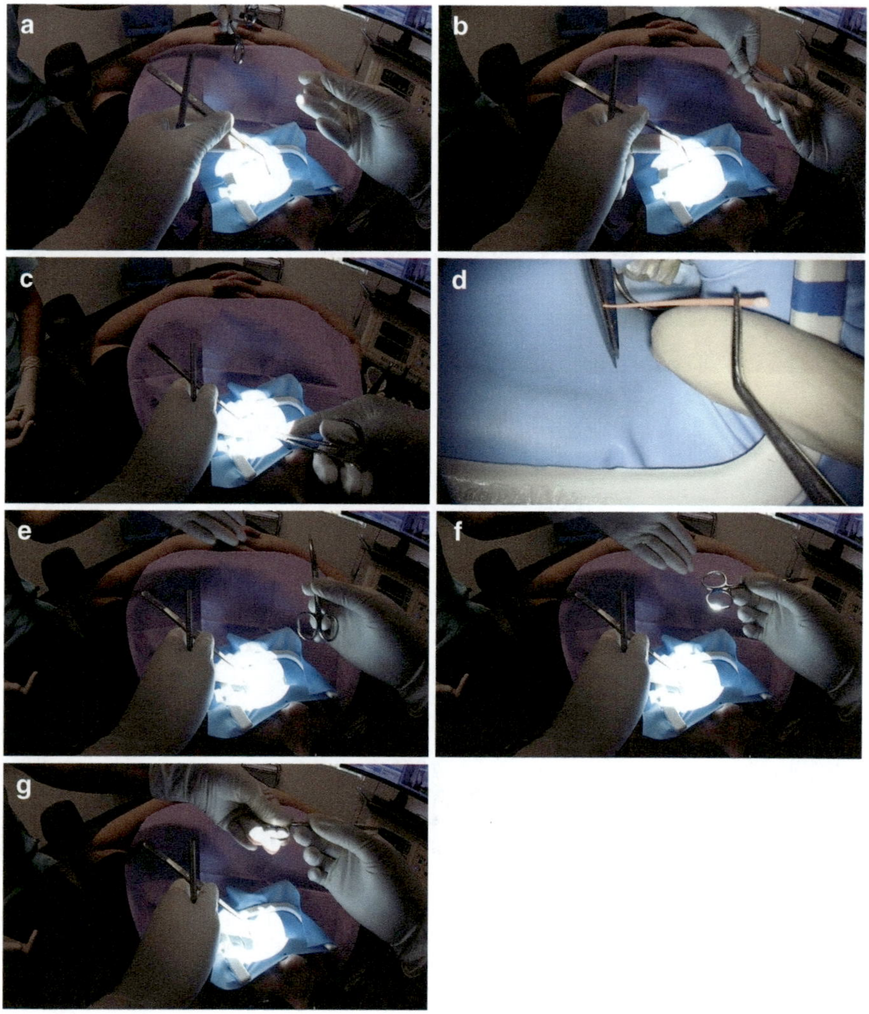

Fig. 5.42 (**a**) Signal to receive scissors, (**b**) scissors being placed with thumb and ring fingers, (**c**, **d**) cutting gutta percha cone in the microscope, (**e**–**g**) returning scissors with an upwards and flick movement with the middle finger

Some methods of fitting gutta percha cones, such as the apical stop philosophy, require the cone size (in relation to the apical stop) to be gauged after rotary preparation. To do this using a precise team approach requires the use of a taper plate. This is a combination of the taper plate being held in the operator's left hand. The tip of the gutta percha cone, held in a locking tweezer, is held in the right hand and placed in the correct size hole in the taper plate. These are held still, and the assistant cuts the tip of the cone with the scalpel. For extra support, the assistant can support the wrist of the cutting hand with the opposite hand.

Suction Handling and Placement

There are different uses for different suction tips. These correspond with why and how they are used. The 90 degree high speed suction is placed as close as possible to the handpiece, in order to prevent the aerosol from reaching the mirror. The assistant holds it in the left hand, so that mirror changes are done using the right hand. This is important in situations such as cutting through metal, where assistant Stropko use may not adequately clear the shards on the mirror, which would require wiping. Another situation is when you are removing a post and require a lot of water cooling with the ultrasonic. The suction is placed close to the tooth and the angle is positioned towards the handpiece, so that aerosol can be removed before it reaches the mirror. For cutting through crowns or restorative material, this is especially valuable. After a mirror change, the assistant cleans the old mirror using gauze with the right hand, whilst looking in the assistant side binoculars and having the left hand supported by the armrest (Fig. 5.43).

The straight surgical suction is used for sodium hypochlorite and EDTA irrigation. For precision, a Luer lock tip (18 gauge, Endo-Eze, Ultradent) is placed on the fine suction tip that can be bent and placed inside the endodontic access if space allows. The tip should be bent in line with the long axis of the tooth and depending on the access size it should be placed touching the pulp chamber floor where possible (Fig. 5.44). For precision placement to occur, the assistant needs to be looking in the assistant side binoculars—otherwise, it will not be placed properly inside the tooth. Stability occurs when the assistant is at rest, which is another reason for having a stool with supporting armrests. All of this prevents aerosol from forming on the mirror. The irrigation suction is held in the right hand because the sponge containing hand files is held in the left hand (Fig. 5.45). During this part of the procedure, there is regular changing between files and irrigants, so this entire process allows for rapid exchange without wasting movements or time.

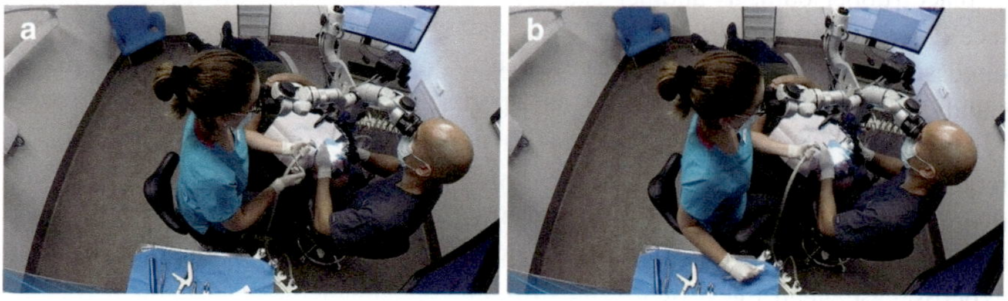

Fig. 5.43 (**a**) Assistant holding high speed suction in left hand and new mirror in right hand, (**b**) mirror cleaning whilst looking in the microscope

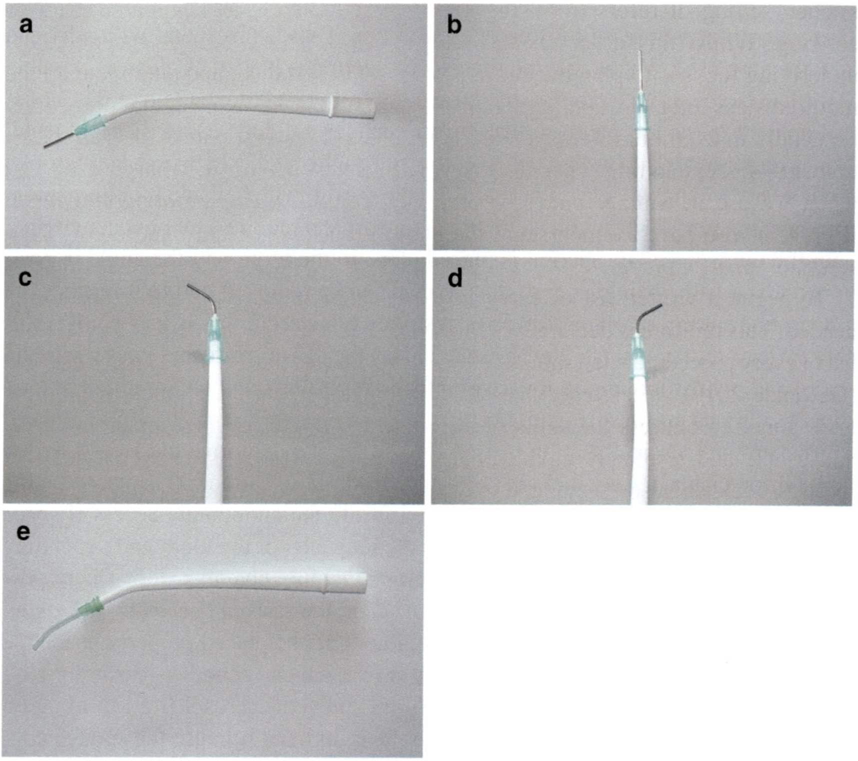

Fig. 5.44 (**a**) 18 gauge Endo-Eze Luer lock tip placed on top of white 1/8″ (3.2 mm) suction tip, (**b**) top view, (**c**, **d**) tip is bent and swivelled so that it can be placed directly into mandibular and maxillary teeth, (**e**) slightly larger gauge SST (Surgical Suction Tip, Ultradent) is used for retreatment so that gutta percha remnants are less likely to clog the tip

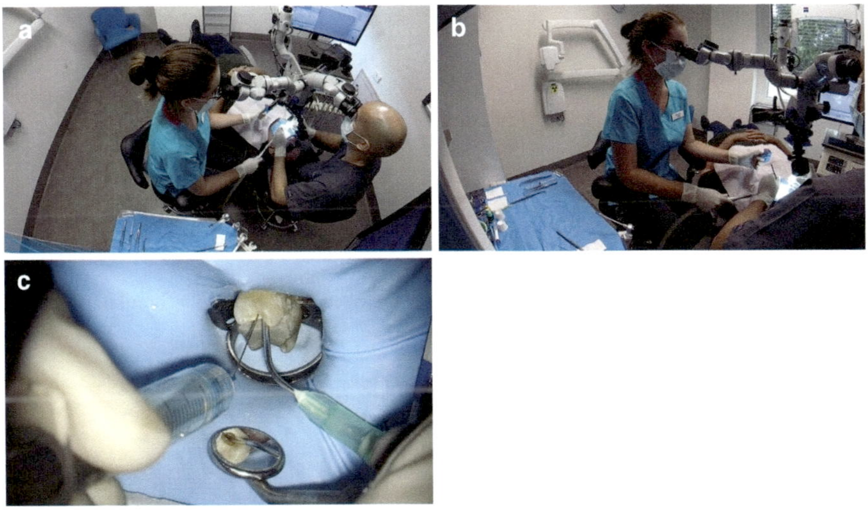

Fig. 5.45 (**a**, **b**) Right hand holding white 1/8″ (3.2 mm) suction with 18 gauge Luer lock tip with left hand holding file sponge, (**c**) Endo-Eze tip on suction tip inside access cavity, preventing aerosolisation and dirtying the mirror

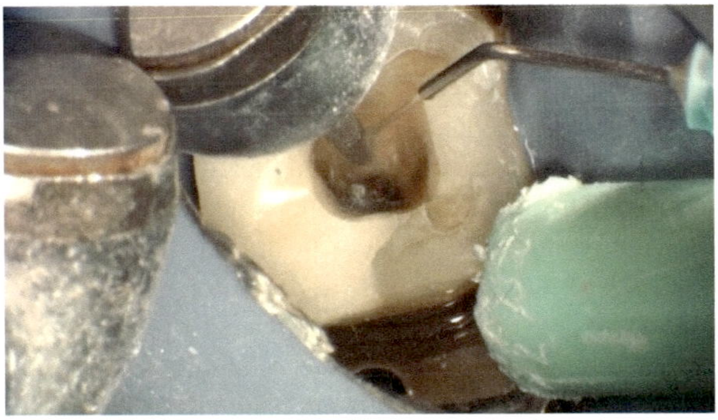

Fig. 5.46 Water stream from assistant Stropko irrigator cooling bur and clearing debris (courtesy Dr. John Khademi)

Assistant Stropko Use

The assistant uses the Stropko irrigator with the high speed suction to manage the operating field. The air/water or air-only Stropko syringe allows for uninterrupted clarity whilst working and is particularly helpful for difficult restorative procedures. Initially, the fine water stream is used to clear the bur (Fig. 5.46). Removing caries and restorative material may often result in debris on the bur, where the water will also cool the bur. The air and water from the Stropko tip is also used to clear the mirror of any debris. Precision placement with the Luer lock tip attached to the Stropko syringe is not only important in correctly aiming for the bur, but also to minimise aerosolisation which causes the mirror to get dirty quickly.

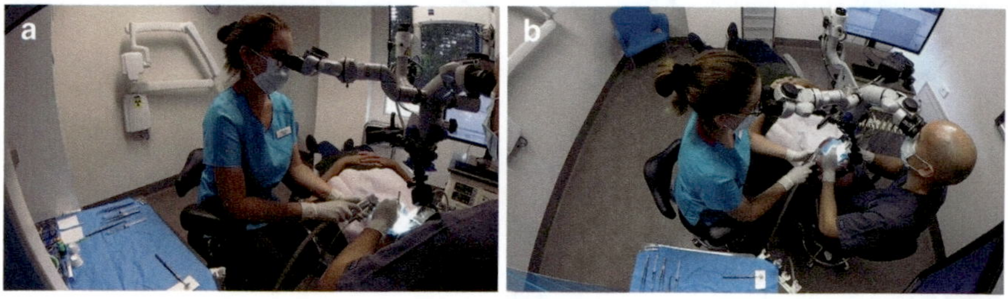

Fig. 5.47 (**a**, **b**) Assistant using right hand to hold 90 degree suction tip with Stropko irrigator in left hand, with both arms supported by armrests

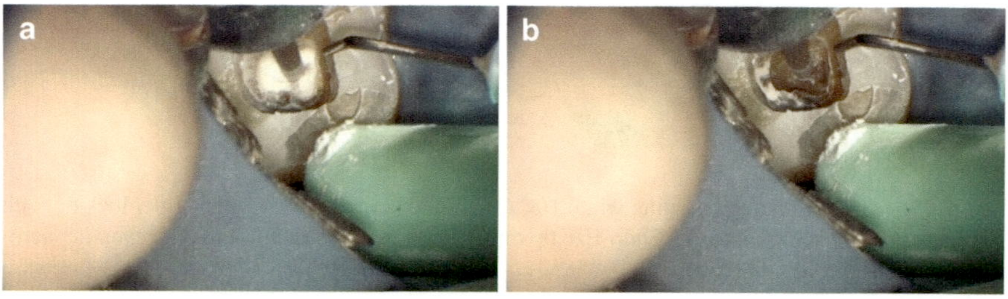

Fig. 5.48 (**a**) Dentine dust impeding operator vision (**b**) operating field immediately cleared of dentine dust using air-only Stropko irrigator (courtesy Dr. John Khademi)

Because of the dexterity required for the precision movement, proper support for the assistant's arms is needed and the operatory should be designed to allow for this to occur. This means there should be enough room for the assistant to sit facing the operating field in the 3 o'clock position with both arms on the armrests—for a right-handed operator (Fig. 5.47).

The 90 degree high speed suction is held in the right hand and when working on the far side—the patient's right—is held lower down on the hose. This gives greater reach whilst keeping the right arm supported on the armrest. The longer 3.5" Stropko irrigator tip is used, which again gives greater reach when working on the patient's right side. For finer exploration, the air-only Stropko irrigator is used, preventing stray water droplets from creating dentine mud and blocking the view (Fig. 5.48). If you only have a single air/water syringe, adding an extension from the air line to become a separate unit is relatively simple and inexpensive. Intermittent air spray clears all of the dentine dust and allows uninterrupted exploration without having to stop drilling, wash, dry, inspect, and repeat—which involves a lot of wasted movement and time. It can even lead to more conservative treatment because the tip of the bur is visualised every step of the way, so dentine removal is minimised. The air/water Stropko cannot be used as an air-only, after water has been delivered, because of the water that is already stuck in the Luer lock tip. This creates mud when mixed with the dentine dust and hinders visualisation. It is important for the assistant to learn both ways—with and without the assistant Stropko irrigator—so that adaptability and expertise can be developed. This is how a team functions at the highest level.

Taking Radiographs

Clear radiographs taken with the rubber dam on can be challenging. In many cases, the microscope can get in the way of a wall-mounted X-ray arm. Moving it to take working length or gutta percha cone fit radiographs can still be an efficient procedure. The bite block is removed to allow slight closure and relaxation of the patient's musculature. The plastic frame is kept on because removing it is not required and is a wasted movement. The tension of the dam stretched over the frame is also an advantage in helping the handle stay in the same position. A swivelling patient chair is required because the arm is often not long enough to image the side opposite to the patient, especially if it is mounted on a side wall. The microscope is moved horizontally, so the focal length stays the same and parfocaling does not need to be redone once the microscope is brought back into position (Fig. 5.49). Where possible, positioning the sensor without the patient holding the handle is preferable because it eliminates the risk of movement—the natural tendency for anyone with an object in the back of their mouth.

First, you need a very good spatial awareness of the position of the X-ray head, the tooth, and the sensor or film. If you know the relative orientation of those three, you are much more likely to get an accurate image and less likely to miss the relevant areas. The main path of the X-ray beam should be 90 degrees to the sensor. Where possible, the long axis of the tooth should be parallel to the sensor. If space is limited, such as a shallow palatal vault, then you want to orient the tooth as parallel to the sensor as possible. For upper anterior teeth, the sensor would be as far back as the molars. For most other teeth, the sensor would be reasonably close to the tooth being imaged.

For the procedure, swivel the patient so the X-ray arm can reach the patient, then tuck the sensor under the rubber dam using the rubber dam clamp as a guide. Imagine you are the X-ray beam casting a shadow. You want to travel right through the tooth, capturing all of the area around the apex, and contain all of your light onto

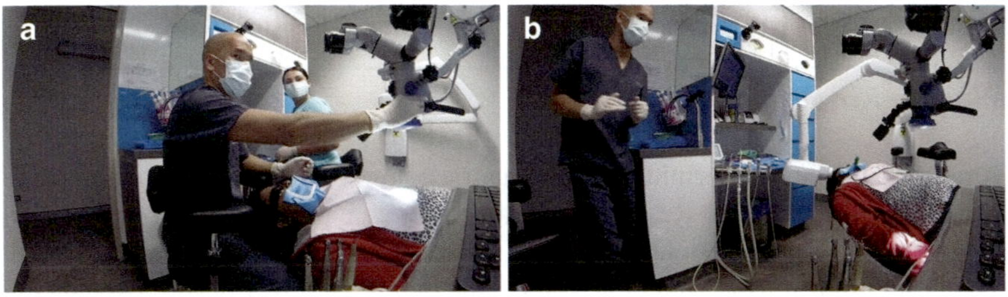

Fig. 5.49 (**a**) Moving microscope horizontally so focal distance stays the same, (**b**) patient chair being swivelled so there is enough room to take right side radiographs

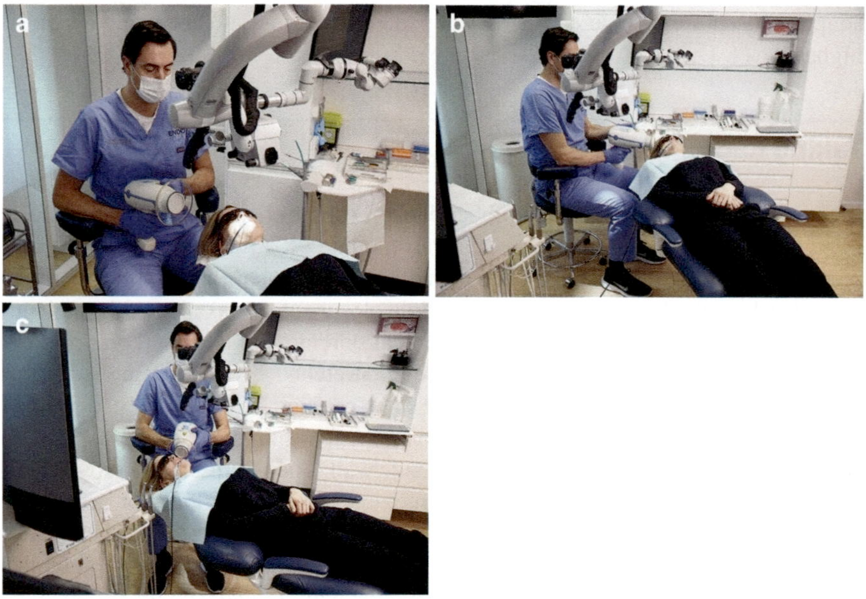

Fig. 5.50 (**a**) Operator getting hand held Nomad portable X-ray machine, (**b**) swivelling patient chair to the left to take right side radiographs, (**c**) swivelling patient chair to the right to take left side radiographs

the sensor. Once it is in the right position, try and leave the sensor as is, without the patient holding the sensor handle. If it moves, you will need the patient to hold it in position. If you have film or a phosphor plate, due to its light weight, you will find that with artery forceps, it will stay in position without the patient holding the forceps, in most cases. This is where the tension of the dam holds it in place. The advantage to the sensor is that you will get the image immediately and can retake with a different angle if required, already knowing exactly which way to change the angle.

In some countries, a hand held X-ray unit is allowed for patient use. If you can use it, swivel the patient chair whilst keeping the microscope in the same position. After positioning the sensor and taking the image, the patient is moved back into the working position (Fig. 5.50).

Placing Amalgam

Amalgam may not be permitted in all countries; however, it is still a great material in certain circumstances. Restorative material placement is one of the most important components of endodontic treatment and a team approach is the most efficient. The best option for the patient is to have the definitive direct restorative material placed immediately after obturation, whilst the patient is still numb and the tooth isolated with rubber dam. In difficult cases where the margins are deep and difficult to access, the microscope is crucial. You will not be able to do high level restorative in these hard cases without a microscope. Quite simply, you will be guessing and the interface between tooth and restorative material will neither be smooth nor cleansable. Not only is this best for the patient, it will streamline the clinical steps for the clinician placing the crown.

In many of these difficult cases, particularly in molars, amalgam is indicated and may be a better option than composite. This is not to say that amalgam should routinely be used, however if you do not have the right skillset to place a complex amalgam build up prior to crown placement, it will be significantly harder for you and your team when it is indicated.

After matrix and wedge placement, check that the matrix band is well adapted and stable by putting a lot of force at the interface with an explorer. The aim is not to get a firm contact, especially in deep subgingival cases. The contact is only left open until the crown is placed, which makes cleaning a lot easier that is essential for soft tissue health. If there is a concavity where you cannot get good matrix band adaptation with a wedge and teflon tape, then you would need to consider an amalgam cutback with the assistant Stropko clearing the field, as shown in the assistant Stropko section.

The quickest and most efficient way is for the amalgam to be placed, then packed immediately. The assistant precisely places the amalgam with the carrier when looking through the assistant side binoculars (Fig. 5.51). Without it, proper visualisation is unlikely to occur, or the assistant would need to bend and lean and put themselves in a potentially dangerous position which is not good for their health.

This starts with the deepest area—in the pulp chamber, a small intra canal plugger (e.g. Buchanan plugger used for packing GP) is used to ensure the amalgam adapts to the deepest part of the pulp chamber. If you are packing under the soffit, or underside of the pulp chamber roof, then a sideways and upwards motion should be used—cleaning the underside beforehand with bent hand files is useful. As you go more coronal, then a larger amalgam condenser or ultrasonic tip that fits in the cavity would be used. This is typically a post removal ultrasonic tip used on low power if any at all—no higher than 25%. Efficiency is important due to the setting of amalgam and as soon as the assistant has placed an increment of amalgam, the next one is prepared whilst the operator is condensing. If the cavity is large, when

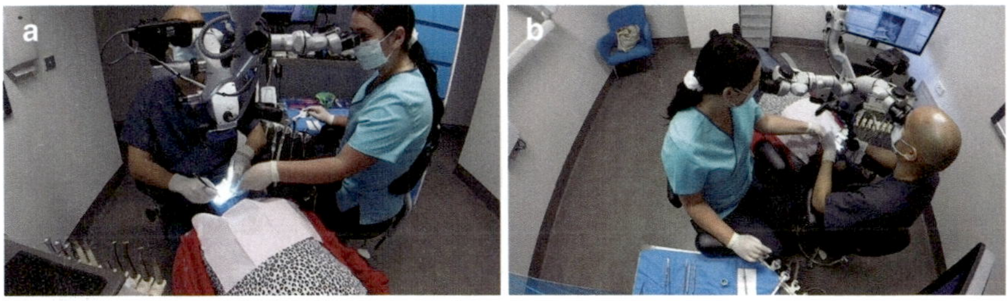

Fig. 5.51 (**a**, **b**) Assistant placing amalgam with left hand whilst looking in microscope

placing the coronal portion of the restoration, consider taking the whole bolus of amalgam with tweezers and condensing the entire amount. The assistant side binoculars allow the assistant to place it precisely so the operator can pack immediately.

Think about what it would be like for the assistant without looking through the microscope and how difficult it would be to have it placed properly. You would be at risk of the amalgam falling out of the cavity if not placed properly, or delaying placement that it sets too quickly (Fig. 5.52).

Full Procedure Simulation

Those of you who have been involved with any aspect of the performing arts will be familiar with the term "full dress rehearsal". This is when you practise the complete performance, in full costume, as if you were there on the night of the concert. All of the individual components that have been rehearsed in their separate sections are combined in the true sequence. If you bought tickets to a show, you would expect a finished and polished performance. As a patient, why would you expect any different from your endodontic or dental procedure? That's right, you wouldn't.

The bigger picture goal of simulating the entire procedure is to *develop expertise for the clinical team*. This consists of two main components: individual skills, plus the ability to adapt.

Experts operate using the recognition of stored patterns. Training for the procedural simulation aims to store all of the patterns required for the procedure, with some variation. In turn, this develops situational awareness, which is a good measure of expertise. An expert assistant will be able to identify the current stage of the procedure and then determine not only what to do immediately next and why it is done, but also what not to do. This starts with the preparation of two key checklists.

The first is a checklist of every instrument and material required for the procedure. This primes the assistant to have every component of the procedure ready, before the patient arrives. If even one step is not prepared for, then the procedure cannot continue. Stopping, then having to find and prepare an expected material or instrument, then resuming the procedure results in a lot of wasted time.

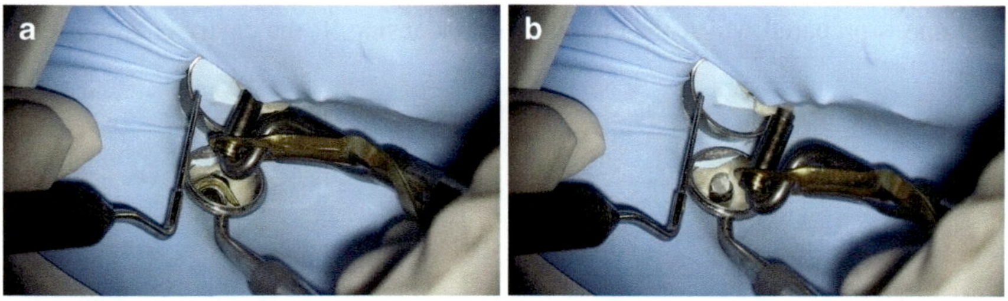

Fig. 5.52 (**a**) Assistant placing amalgam carrier in access cavity before placement, (**b**) immediately after placement, operator is ready to condense amalgam

The second is a sequential list of every step in the procedure, from when the patient sits down to when they leave the chair. Logically, there should be one order in which the required tasks occur. Understanding why they occur is also important—i.e. determining the canal lengths occurs before the rotary instrumentation because the operator needs to know first how far to prepare the canals. This list should be easily referred to in the operatory so that there is no confusion. These steps should occur with the least amount of wasted movements. As a result, procedural knowledge is attained which is valuable for all team members to perform their portion of the procedure. Readers interested in the benefit of a checklist should read the book The Checklist Manifesto by Atul Gawande.

Sample Start Treatment Setup and Procedure List

Before patient arrives—setup
Check tooth number and treatment type. If retreatment, get chloroform (solvent).
Handpieces—high speed: round diamond, rounded tapered fissure diamond
slow speed: Lentulo spiral, round bur if caries removal needed
Ultrasonic tips
Stropko tips
Treatment cassette—instruments out and on bench
Matrix band to suit tooth, wedge (check pre-op PA)
File pack (new #10 and #15 K files) with sponge and ring
NaOCl with needle bent
Suction tips: green, then white with Endo-Eze tip attachment
XDR sensor with ComfortWand holder
RD with clamp and frame
LA syringe setup as per tooth, with topical LA on cotton bud
Rotary handpiece
Apex locator cords
Cavit, corner of file sponge removed
Ultracal
Temporary material (Fuji IX, Fuji II LC or flowable composite)
Articulating paper
Gauze
Floss

Patient seated in chair
Bib and glasses
Mirror
Topical, then LA syringe
Saliva ejector during LA administration
In "Treatment" tab, complete Anaesthesia entry, irrigation protocol
Working mirror and bite block

RD, floss
Parfocal
Green suction to remove restoration, second mirror in right hand (gauze on bench to clean)
Fine suction with Endo-Eze green tip for irrigation inside canal
Wash
Long #10
Enter treatment page details
Different hand files as per case
Apex locator with long #10 file
Take PA ("WL File") in Acquire window
Rotary prep
Ultracal, Cavit (damp sponge)
Restorative material (Fuji IX, II LC, or flowable composite)
Take RD off, suction and dry
Articulating paper

Sample Complete Treatment Setup and Procedure List
Before patient arrives—setup
Check tooth number and treatment type. If retreatment, get chloroform (solvent).
Handpieces—high speed: round diamond, rounded tapered fissure diamond slow speed: Lentulo spiral, round bur if caries removal needed
Ultrasonic tips
Stropko tips
Treatment cassette—instruments out and on bench
Matrix band to suit tooth, wedge (check pre-op PA)
Patient file pack with sponge and ring
NaOCl with needle bent
EDTA with needle bent
Suction tips: green, then white with Endo-Eze tip attachment
XDR sensor with ComfortWand holder and Rinn post-op PA holder
RD with clamp and frame
LA syringe setup as per tooth, with topical LA on cotton bud
Sealer
GP cones in measuring block
Paper points—2x per canal
System B tip and Elements cartridge
Etch with microbrush
Restorative material setup: If Luxacore = Centrix tip and syringe, photo bond and microbrush, dry microbrush, fibre post.

If Amalgam = amalgam, instruments, damp cotton bud
Articulating paper
Gauze
Floss

Patient seated in chair
Bib and glasses
Mirror and back of perio probe for percussion
Topical, then LA syringe
Mix sealer during LA administration, saliva ejector
In "Treatment" tab, complete Anaesthesia entry, treatment details
Working mirror and bite block
RD, floss
Parfocal
Green suction to remove restoration
Fine suction with Endo-Eze green tip for irrigation inside canal
Wash, #15 as per WLs [short: all canals 21 mm or shorter, long: one canal 21.5 mm or longer]
Wash, MAF rotary files by hand (top to bottom)
Wash, GP cones (top to bottom)
Cone fit PA
EDTA, wash, paper points
If post, fit post and record length for canal
Sealer, GP cones, burn with System B one at a time
Downpack with small Buchanan, one canal at a time
Backfill with warm GP, left hand tweezers
Etch, damp microbrush
Check PA
Remove rest of temp (green suction)
Matrix and explorer, wedge(s)
Restorative material and polishTake RD off, suction, articulating paper
Post-op PA

Surgical Training

Team training for endodontic surgery is very different because two assistants are involved in the procedure, one looking in the microscope and managing the field (left side of the operator), the other sitting on the opposite side managing instruments and materials. For a right-handed operator, instrument exchanges are done with the assistant on the right side. See the surgical chapter for more details.

Team Footage Review

Recording and visually reviewing your workflow is valuable if you are aware of what to focus your attention on and how to change. For all motor and procedural skills, how someone believes they are moving or behaving can be very different to what is actually occurring. Even if you are training in front of the mirror the entire time, your attention would be largely focused on the task at hand. Video recording to evaluate your movement and behaviour is used in many other domains at elite level—for example, professional sport, performing arts. At a team level, analysing how one or more people interact is just as important as monitoring individual performance.

Visual evidence can influence and even reverse one's perceptual beliefs of behaviour.

In the past, video recording and review technologies were expensive, slow, and impractical.

Now, not only are basic digital recording devices cheap and quick to allow for review, but more advanced technologies like slow motion and high resolution video are now inexpensive and accessible to the consumer. The general approach is to identify the individual ergonomic strengths and weaknesses, such as positioning and movement, build on and correct them.

Basic principles about being at rest, having a neutral and relaxed position and moving the patient instead of the microscope, can be easily noticed. Weaknesses can be described by referring to the classes of motions and trying to have less wasted movement where possible. Ideally movements should be done using a class 1 or 2 motion. If something can be done with a class 1 movement (moving the fingers) and the operator is bending at the waist with a class 5 movement to reach for a hand instrument, then the team approach would be to have the assistant pass the instrument into the operator's hand. If a handpiece is being taken with a class 5 motion, then keeping the elbow on the armrest with a class 3 motion is the way to resolve the situation. Acknowledging what has been achieved, and what the goal for the upcoming treatment sessions are, should be clearly defined.

The overlying culture should be reinforced during team analysis of video recording. In this book, it is centred around the health and safety of the operator and assistant, to minimise wasted movements and increase efficiency.

Progress appraisal, required milestones

Focusing on a specific goal, corresponding to an issue identified in a previous review of footage, is an effective way of improving rapidly. Both training and treatment procedures with patients should be recorded for review. Training sequences for both individual skills and procedural simulations can be analysed, then reviewed for improvement. The role reversal of the assistant sitting as the operator is valuable to analyse because the assistant can gain a better understanding of not just the operator's role in the team, but with video playback, how improvements can be made to help the team function together to the highest level (Fig. 5.53).

Before a review, both operator and assistant should note down what they initially perceive to be ergonomic and efficiency issues, and identify aspects of the treatment

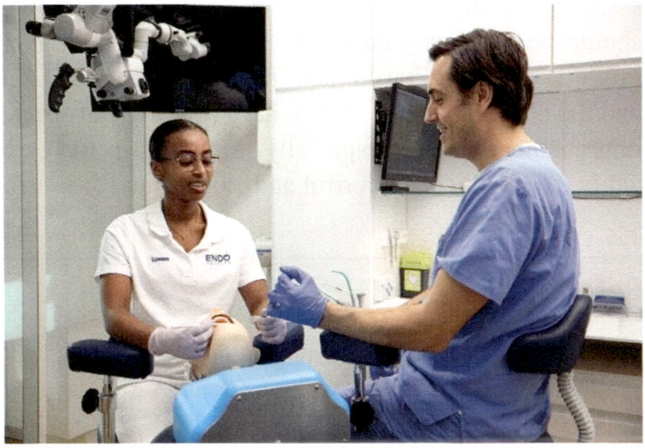

Fig. 5.53 Operator and assistant training by switching positions in order to gain a better understanding of each role

that could be improved. Milestones should also be set, for example, in 1 day, class 5 movements will be eliminated, in 2 days, class 4 movements will be eliminated, in 3 days, class 3 movements will be minimised, and so on.

During and after the review, three key points for each team member to improve on should be established, based on what has been identified. If this is consistent with the original issues raised by each member, it indicates some level of situational awareness, a positive sign in skillset development.

If not, then there may be more than three key points to improve on.

During initial learning stages, re-training without patients to refresh and further develop skills should be done more frequently. This should be done before a conscious effort during the patient procedure to implement changes.

Focusing on goals one at a time, in order of positioning and movement importance should be done. For example, if you notice that the operator is slouching in the chair, then sitting up straighter with a more neutral spine should be the first focus, instead of say, how the irrigation syringe is aspirated. If there are many class 5 movements, then limiting them—by the operator not reaching over to take instruments or materials—should be the focus. The progressive step is to then limit class 4 and 3 movements. Due to the assistant having two work zones—the back wall tray for instruments and the operating field, there will be more class 3 and 4 movements. However, there should be focus on neutral positioning and using the rotational stool movement with the feet—to ensure that there is no leaning or twisting at the hip, which can lead to physical strain or injury.

Certain training drills or modifications can also be implemented. If the assistant or operator is having their eyes too close to the eyepieces, a feather can be taped to the binoculars as a deterrent. If the operator is lifting their elbow above the armrest too often, then a string or elastic band can be used to tie the operator's elbows down. These training drills are to develop good habits and get rid of bad habits. If the right-handed assistant is passing hand instruments with the right hand, hitting the mirror and disrupting the field of view, training with the right hand held behind the back can be done.

This develops muscle memory and automation.

At the subsequent review, all identified issues from the previous video recording should be first analysed to check for changes, then any new perceived issues are raised and then reviewed, just like in the first video. Clear goals and expectations, and the ability to assess if they have (or have not) been achieved, can lead to quick improvement. This is especially important in the early stages of training, where deliberate practice is effortful and can produce powerful results.

References

Hilgard ER, Marquis DG. The definition of learning. In: Kimble GA, editor. Hilgard and Marquis' conditioning and learning. Appleton-Century-Crofts, Inc; 1961. p. 1–13.

Horn J, Masunaga H. A merging theory of expertise and intelligence. In: Ericsson K, et al., editors. The Cambridge handbook of expertise and expert performance. Cambridge University Press; 2006. p. 587–611.

Clinical Microscope Endodontics: Consultation and Case Presentation

6

Abstract

This chapter will discuss the important features of a consultation appointment and how an ergonomic workflow incorporates these factors. Communication topics and their relevance to patient understanding are outlined. Assistant and operator positioning and the importance of a starting protocol are covered.

Key Points

- Entire operatory setup should be completed before patient arrival.
- The first impression can influence the entire consultation experience. Building rapport with conversation is important in helping the patient feel comfortable.
- Case presentation should occur with the operator directly facing the patient, so that all details can be discussed openly, and the patient has confidence in the management plan.

When patient appointment components are structured from a team approach, a logical way to break down the preparation is into three parts: before, during, and after the appointment. This simplifies the understanding of each team member's requirements. It also shows the order in which sections should be prioritised.

K. E. Yapp, C. Aznar Portoles, *Team Microscope Endodontics*,
https://doi.org/10.1007/978-3-031-70188-7_6

Before the Appointment

The use of a checklist is helpful with implementing a systematic and well-organised work environment. Checklists are used as cognitive aids to guide users through task completion. In medicine, they can be an effective tool to improve care processes and reduce mortality and morbidity. In the endodontic office, especially when setting up a team or when teaching any new member of staff, they can be very efficient to create work routines and avoid mistakes during consultation/treatment.

Setup and Checklist

All aspects of the operatory should be setup prior to the patient entering. This includes the overall cleanliness, instruments/materials on the back wall tray, dental cart with air/water syringes, orientation of the patient chair (locked into position) so that the seated patient will face the operator when entering the room, patient file opened in the practice management software and the music playing (Figs. 6.1 and 6.2). Whilst the

Fig. 6.1 Patient chair locked and aligned in the correct orientation

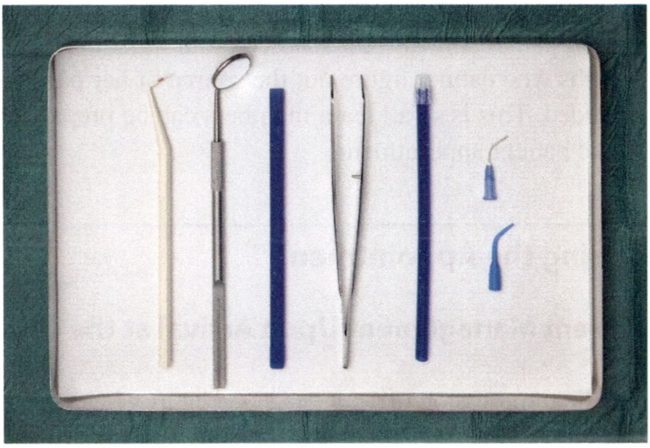

Fig. 6.2 Consultation instrument setup

patient is in the room, setup should not occur "on the fly"—if all of the preparation is not completed, that will lead to unnecessary inefficiency and wasted time during the patient appointment.

A printed checklist should be attached to the wall, for easy reference, so that no detail is missed.

Example Consultation Instrument List
Mirror
Periodontal probe
Sickle explorer/Briault probe
Tweezers
Air/water syringe (or Stropko syringe)—on dental cart
Other instruments to have available:
Articulating paper
Pulp test (cold test)
Fracfinder (Tooth sleuth)

In a referral practice, most of the information is usually received before the patient arrives at the office. History of previous treatment, recent and historical radiographs or a brief description of the presenting problem will help with getting the consultation/treatment room ready. In an ideal clinical setup, all the different steps of the treatment, from consultation to surgical or non-surgical treatment, are performed to the best standards and without interruptions that might affect the normal workflow of the daily schedule. Many clinicians dedicate different rooms for different treatments. This is not only inefficient but it also increases the overhead costs of the office. The design of the workspace must be adapted to the different needs of the team and the patient.

Any other relevant patient information should be collected by the reception staff prior to patient arrival. Details such as the patient's expectations (e.g. "Mrs X is travelling from three hours away and would like to have treatment started today"), practical considerations ("Mr Y uses a wheelchair and is being driven by his carer to the appointment"), or even details about the patient history (e.g. "Ms Z wants to get her pain resolved—she has seen four other dentists and specialists in the last six months who cannot figure out the source of her pain") are important and should be recorded. This is so all team members can be prepared and can adapt for all aspects of the patient appointment.

During the Appointment

Patient Management Upon Arrival at the Office

The first impression of your practice begins when the referring dentist mentions your name, and when the patient looks at your website/business stationery. The patient's first contact with the office takes place at reception. Physical aspects of the entrance and waiting room can affect the patient experience. Is it too cold or too hot, relative to the outside temperature? Does it feel warm and inviting? Is there any music, external noise, or is it completely quiet? Was the office difficult to find regarding transport, parking, and signage? All of these factors can change the patient's attitude before they set foot into the practice (Fig. 6.3). The task of a receptionist is not just to welcome the patient, but to try to get all the possible information that might be relevant for the clinical team that has not been collected already: How is the patient feeling today? Is he/she nervous? Is the patient alone or accompanied by someone? A medical history form is completed and signed before the patient enters the room. If the patient has any questions, the receptionist should be trained with helping out with this step. Patients can often forget important medical aspects or medicines that may be relevant for the treatment.

First Meeting with the Assistant

The first impression that the patient gets when entering the room can influence the rest of the consultation. The setup needs to be organised and consistent. The assistant welcomes the patient and helps him or her to feel comfortable and establishes rapport with conversation. The goal of this first conversation is to get to know the

Fig. 6.3 Waiting room furnished with a warm and pleasant décor

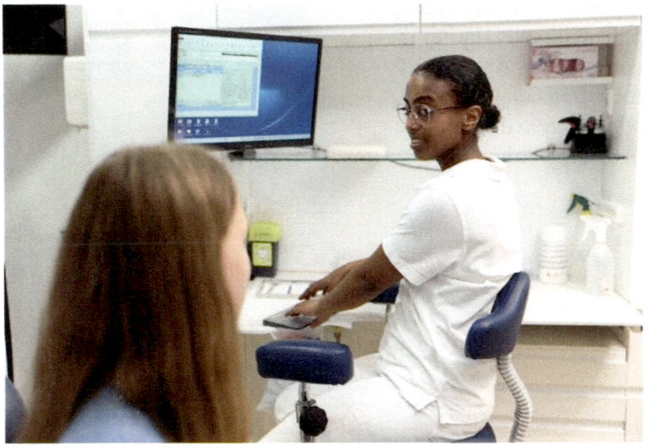

Fig. 6.4 Assistant talking with patient and typing information regarding clinical history into patient record

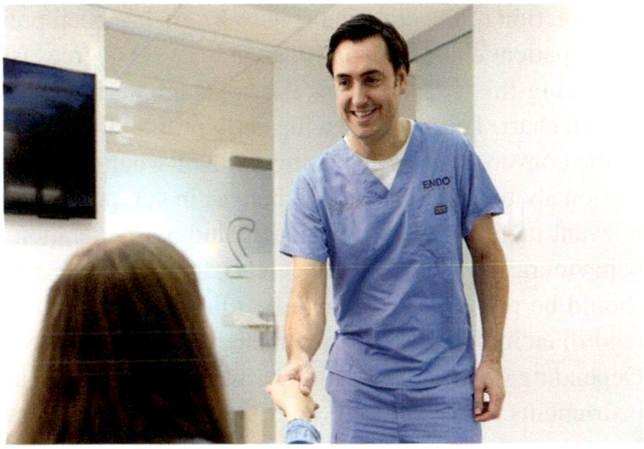

Fig. 6.5 Operator greeting patient directly and front on

patient and collect other non-clinical information (Fig. 6.4). This information is entered in the assistant side computer. At this stage, the operator is still not present in the room. This initial meeting can be very helpful for patients suffering from anxiety. The assistant should be trained on effective communication with the patient, should listen to them and empathise with their fears or worries.

First Meeting with the Operator

The operator enters the room and faces the patient directly (Fig. 6.5). A friendly and professional attitude is important, as it helps with building rapport. Because all the previous information has been already collected by the receptionist and by the assistant, the clinician will proceed to let the patient explain "in his or her own words" what the presenting issue is. During this conversation, the operator will focus on the issue from the patient's perspective.

Any additional information given by the patient can be valuable, and it can also help with establishing a bond with the clinician. Mutual hobbies, interests, etc. that identify a common ancestor can help with building rapport and building trust to a more personal level. The conversation takes place with the patient chair turned in

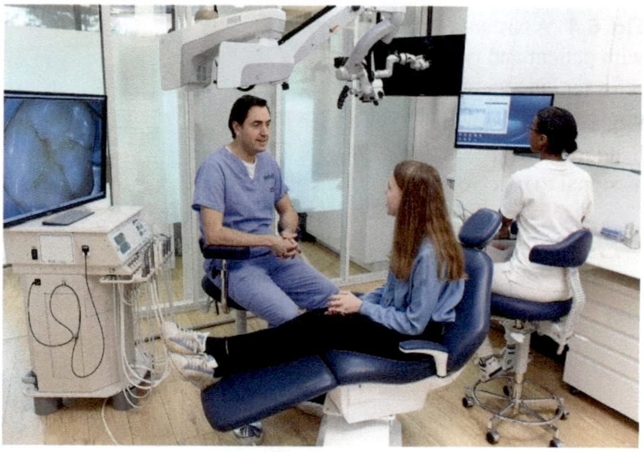

Fig. 6.6 Operator facing patient directly during consultation. Assistant is recording details of conversation directly into patient record

the direction of the clinician (Fig. 6.6). As has been previously emphasised, a swivelling patient chair that can rotate at least 60 degrees is crucial for this purpose.

During this conversation, the assistant starts entering the information into the patient chart. The clinician will subtly prompt the assistant to enter particular details of the conversation by repeating parts of what the patient says. In particular, information about the pain history, changes in symptoms, previous treatment performed, relevant medical history details (including medication taken), and relevant patient behavioural details should be recorded. Previous radiographs from the referrer should be present, so data can be filled in the diagnosis component of the patient record: including previous treatment, caries, restorations, radiographic findings, etc. Depending on the conversation between the operator and the patient, any additional instruments that would not be routinely used (e.g. Fracfinder) can be gathered in addition to the regular consultation instruments already setup.

In order for the assistant to be able to effortlessly enter this data into the patient's record, the computer needs to be located in the circle of influence. This computer is usually located within the back wall at the closest end near the operator. At this position, the assistant is able to face this computer directly, take notes or clinical pictures, and exchange instruments efficiently.

After this introductory conversation, the operator will proceed to examine the patient. The patient is informed about the tests that will be conducted during the examination and their purpose.

Starting Protocol and Patient Examination

The operator and patient chair presets should be setup prior to the patient entering the room. The operator stool height should be adjusted so that the feet are flat on the floor, and the knees and hips are at a 90 degree angle. The operator should be sitting with a neutral spine and the torso upright. A neutral spine is when all three normal curvatures of the spine—cervical (neck), thoracic (middle), and lumbar (lower)—are present. This will determine the armrest height, which should have the elbows by the side of the torso and the shoulders relaxed. Too high and the shoulders will

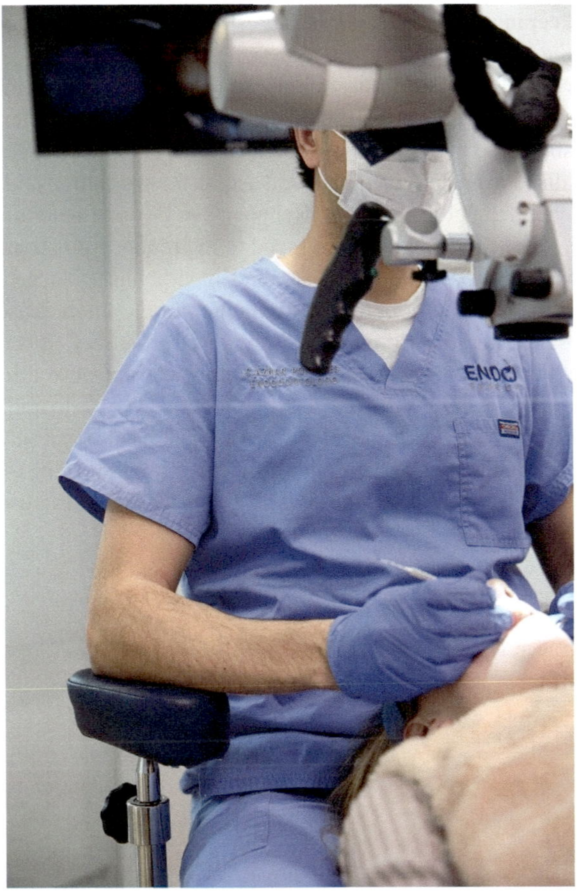

Fig. 6.7 Operator with neutral spine and arms on armrests

be shrugged and under tension, too low and the forearms will be unsupported. The backrest height should be set so there is adequate lumbar spine (lower back) support. The forearms should be parallel to the floor and the armrests supporting this position (Fig. 6.7). With the hands in this position with forearms on the armrests, this determines the height of the patient chair.

The patient chair is reclined so that it is flat and parallel to the floor. The height is adjusted so that the patient's mouth is slightly below the level of the hands. This preset will be consistent with a majority of patients because the size of the adult human head has a range of normal. If the head is too high, the operator's arms will be angled upwards, leading to continued strain and a lack of arm support. If the head is too low, the hands will not be able to reach the mouth and the arms will have to be taken off the armrests. In this position, the knees will be on either side of the patient chair for most people—which allows the legs to swivel the patient chair side to side. For tall people with long torsos, the legs could be underneath the patient chair, which is fine—as long as the patient chair can be swivelled with the legs.

The dental cart with air/water syringes and handpieces is moved to a position where they can be reached by pivoting at the elbow, a class III movement, whilst keeping the elbow on the armrest. This prevents inefficient reaching and moving at the shoulder and hip.

The assistant stool should also be adjusted in the same way as the operator stool, with armrests supporting the forearms. The main difference is that the stool height should be higher than the operator. This is so that a greater overall view can be visualised from above, unimpeded access for suctioning and Stropko irrigator use can be obtained and instrument exchange can be done without raising the arms too high.

The patient chair is changed to the reclined position and set parallel to the floor, with the occlusal maxillary and mandibular planes being bisected by a line 90 degrees to the floor. A chair with memory foam increases patient comfort and helps with tolerating longer procedures. If necessary, the operator makes any patient adjustments to maintain an optimum work posture throughout the entire consultation. This might include asking the patient to slide up (towards the operator) in the chair, if they are slouched or short. The head position should have the top third of the patient's head hanging off the edge of the headrest (Fig. 6.8). Conceptually, the headrest should be considered a "neck rest"—because the neck is the important

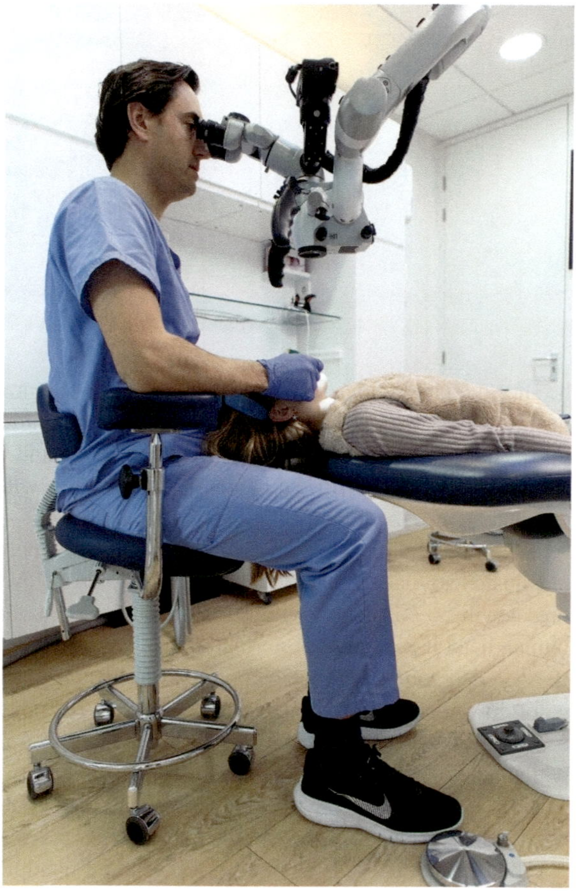

Fig. 6.8 Patient neck supported by neck rest with top third of the patient's head hanging off the edge

body part to be supported. The working position of the operator should remain always constant; any necessary adjustments will involve swivelling the patient chair and changing their head angle. The microscope is then brought into the field, with the body of the microscope angled at 90 degrees to the floor. The long axis of the lens direction is parallel with the torso and enables the operator to sit upright and relaxed without a hunched position.

It is particularly important that the scope is parfocaled before starting the consultation to be able to take good quality photography or video; otherwise, the camera will not be in focus when the operator has a clear field. The object of a parfocal exercise is to get the microscope eyepieces and the camera to focus at the same distance. This is covered in Chap. 3.

Once the operator has finalised the microscope position and completed the parfocal adjustment, a signal is given to the assistant by putting both hands on the microscope, who will make any required adjustments to their working position. The binocular tube angle is adjusted, then the diopters, and finally the image rotation adjustment is modified to observe the image in the correct orientation (3–4 o'clock vs 12 o'clock operator position) (Fig. 6.9).

This *starting protocol* should be done every time a patient is reclined, regardless of the appointment type (consultation, review, treatment). The entire protocol of checking the patient position in the chair, bringing the microscope into the right position and making adjustments for operator and assistant will ensure that a consistent working process occurs. It prevents wasted movement by having to move the microscope during treatment.

The first part of the examination is conducted with the microscope at its lowest magnification with the widest field of view. This will aid performing an extraoral examination, checking for potential signs of pathosis or facial asymmetry. Lymph nodes and thyroid are also checked and oral cancer screening is performed. The patient should perceive the extra value of the examination process, where the operator does not just focus on the problem that the patient is referred for. Low magnification movements are subtle because of the wide field of view. To do this, the patient's head is turned slightly in one direction. This is not tilting the head where the ear is moved closer to the shoulder, instead, it is identical to when one crosses the street and looks left and right for oncoming traffic. As soon as the head is turned, the area of interest is moved outside of the light and outside the field of view. How would you compensate for this? The most common error for an inexperienced and/or untrained operator is to move the microscope. This is wrong because of the constant and excessive wasted movement. The correct way is to swivel the patient chair in the opposite direction using the legs. This is the only way to work efficiently for both operator and assistant. This basic principle is never used in naked eye dentistry because the swivelling patient chair is rarely used, and there is often a tendency to adjust the overhead chair light.

During the second part of the consultation, special attention is given to the TMJ, occlusion, and the whole dentition. Excursive and protrusive movements to check for occlusal relationships or interferences can be examined easily due to the wide field of view. The periodontal situation is assessed and restoration margins are

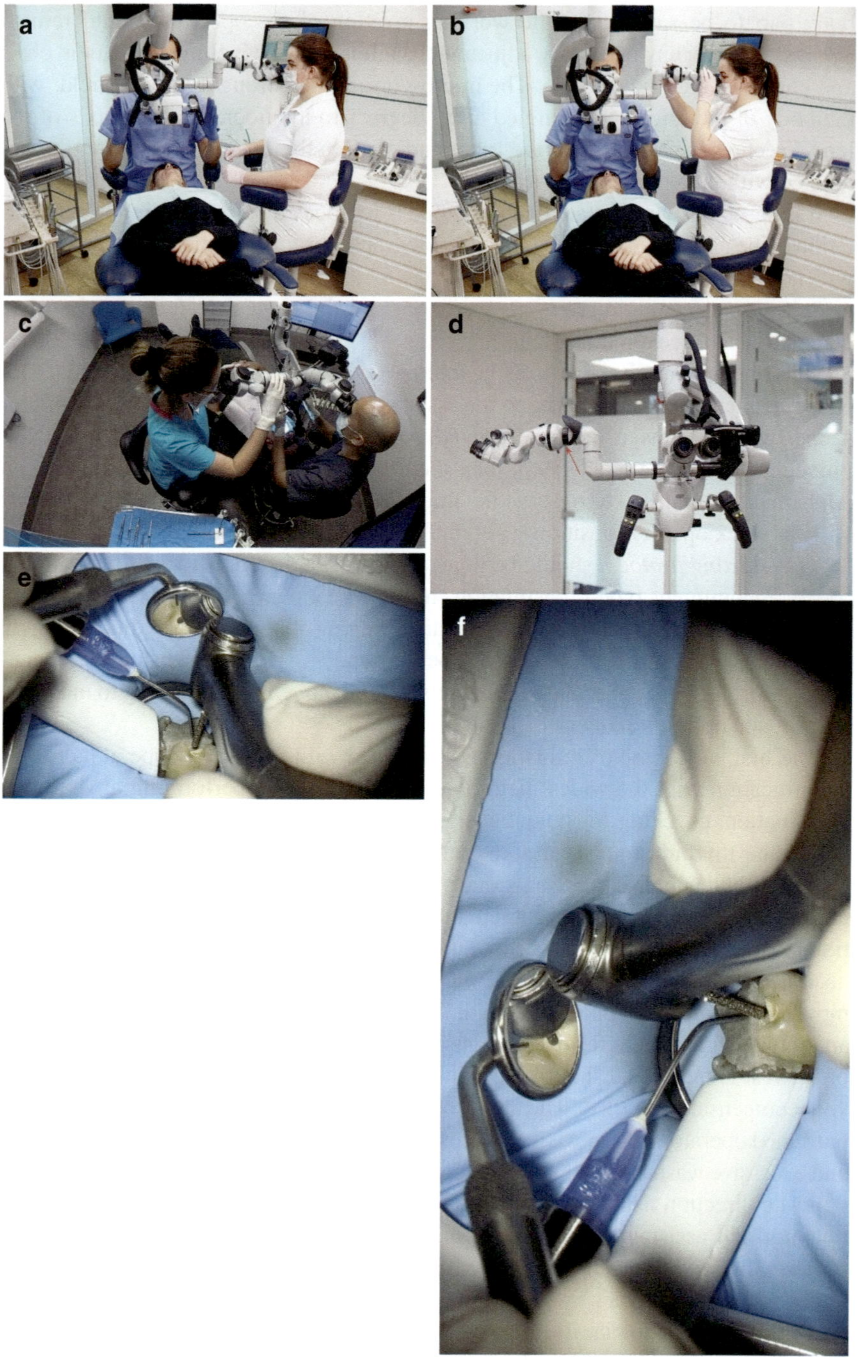

Fig. 6.9 (**a**) Operator ready to signal to the assistant the parfocal is complete, (**b**) operator securing microscope with both hands and assistant adjusting binocular position, (**c**) assistant adjusting image rotation ring, (**d**) red arrow showing image rotation ring on assistant binoculars, (**e**) operator view through the microscope at 12 o'clock position, (**f**) assistant microscope view at 3 o'clock position should be rotated to be approximately this orientation

examined under the microscope. Air from the triplex or Stropko syringe is used to dry the field and obtain a clear view. The assistant observes the examination looking into the assistant binoculars and is able to contemporaneously enter the details into the patient's record as findings are noted. Once these additional examinations are completed, the operator will focus on the problem tooth or teeth.

The different tests will be performed sequentially, following a structured routine. This sequence should not be altered, as this would introduce uncertainty, and efficiency would be lost. An example of a sequence of tests is:

- Probing depths
- Tooth mobility
- Percussion
- Palpation
- Pulp test (Cold/Heat)
- Bite test (Fracfinder)
- Radiographs

This order is chosen because the periodontal probe is inverted 180 degrees by the operator to use its back end to check mobility and percussion. To avoid another switch, once finished with the back end, it is passed back to the assistant.

Intraoral Imaging and Documentation

Documentation is also performed contemporaneously. With modern DSLR and mirrorless cameras, taking good quality photos can be done effortlessly. Intraoral pictures are normally taken by the assistant and should be automatically stored into the patient's chart as they are taken. Integration of the camera program and patient software is therefore mandatory. The assistant position should provide a direct vision of the operator's cart monitor, as a second wireless mouse will be used to trigger the camera remotely. The operator verbally indicates to the assistant when a picture has to be taken or the assistant sees through the binoculars that the object is in focus, the mirror or patient is still, then takes the photo (Fig. 6.10).

Radiographs are usually taken during consultation for diagnosis, treatment planning, and follow-up evaluation. In order to take radiographs efficiently, a few aspects need to be taken into account. When designing an endodontic operatory, the main focus is the microscope position. This will be dependent on multiple factors, like proximity of walls, height of the room ceilings, or other building factors. The type of X-ray unit employed has a direct influence on the position of the microscope and the position of the patient chair. If a fixed radiograph arm is going to be installed, special attention has to be given to the length of the arm in relation to the head of the patient chair and the position of the microscope, as they can get in the way when in use. In order to be able to efficiently take radiographs from all quadrants, the chair will be swivelled so the X-ray arm can reach all the quadrants being treated.

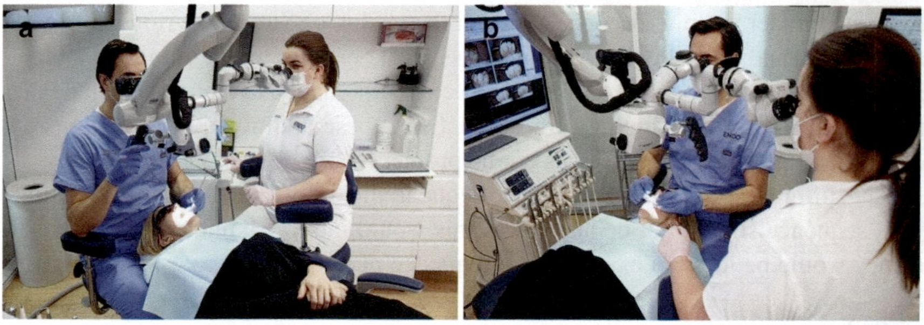

Fig. 6.10 (**a**) Assistant using mouse to take photograph, (**b**) assistant looking at monitor to ensure cursor is in the correct position to take photograph

When a portable radiographic device is employed, there are less distance and spatial constraints. The key point to remember is that, when not being used, the device should be located within the circle of influence of the operator or from the assistant, if the assistant is the one handing it to the operator.

In contemporary endodontics, digital radiographic systems are mostly employed. The main advantage of these systems, as compared to analogue ones, is that they produce a better-quality image with less radiation. Other advantages are the elimination of chemical processors and that the need to dedicate space for cabinets of analogue images is not required.

The two main digital dental radiographic systems are sensors and phosphor plates. Both systems present different advantages and disadvantages.

In the endodontic operatory, a digital sensor is preferred because of its efficiency (faster image acquisition than phosphor plates and fewer motions involved—no radiograph development is required). The sensor is plugged via USB cable to the computer located at the cart (Fig. 6.11). When not being used, the sensor should be placed in a position which is easily accessible for the clinician in the circle of influence. Modern endodontic carts are equipped with external USB connections, allowing the clinician to easily connect the sensor without having to open the cart, if the PC is located inside.

Phosphor plates are an accurate image acquisition system. Their main advantage is that they can be used in multiple operatories and that the plates are easily replaced when damaged. They are also relatively cheaper than digital sensors. The disadvantages are being less efficient than digital sensors as they involve more steps and processing time: the operator or the assistant will have to develop the plate before the image is acquired and leads to a longer waiting time. If a phosphor plate is going to be used in the endodontic office, the ideal location for the scanning station is inside the operatory, as in this position the operator or the assistant will need to leave the room to develop the phosphor plate. In an ideal setup, this scanning station should be located within the circle of influence of the assistant, so that this image can be efficiently developed immediately after it is taken.

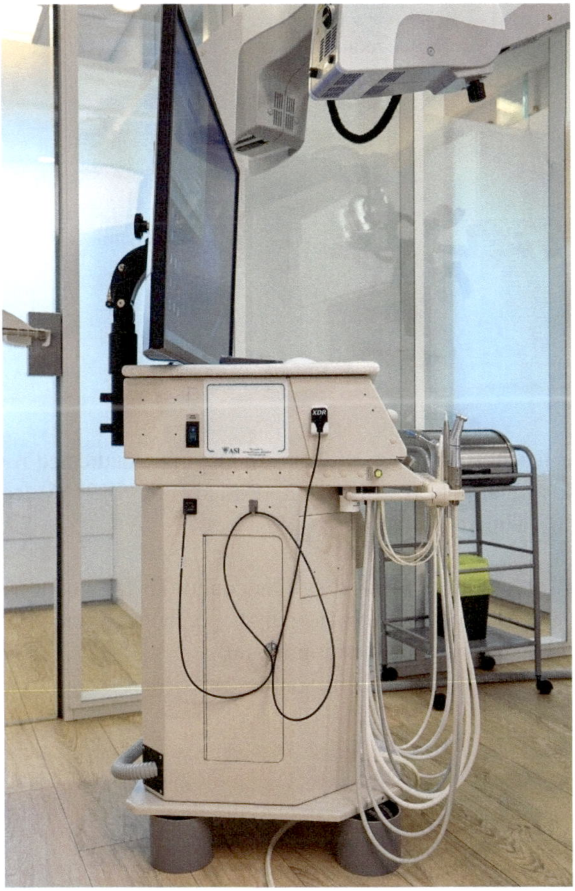

Fig. 6.11 Dental cart with USB connection for digital X-ray sensor

Taking good quality intraoral images and radiographs enhances the communication with the patient and with the referring operators and are a powerful tool for teaching purposes.

CBCT Study Review

CBCT imaging is nowadays routinely used in the endodontic office for diagnostic purposes and treatment planning. Thought should be given to the location of the CBCT unit in the office. Ideally, patient walking distance and travel time should be kept to a minimum. A CBCT room located close to the operatory is most efficient as it reduces this time when a CBCT study is indicated.

The radiographic study may be taken prior to seeing the patient or after the consultation, if indicated. A preliminary read of the CBCT study can be taken with the patient chairside when intended for patient education or for case presentation, but the characteristics of an operatory usually make it unsuitable for performing an

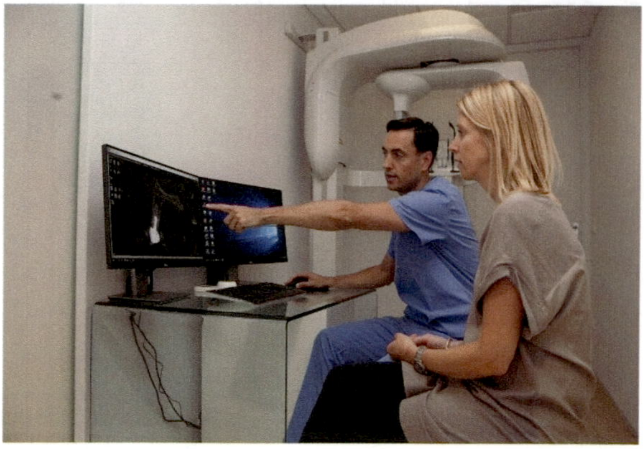

Fig. 6.12 CBCT review in a separate, dedicated room

efficient read of a CBCT study. Having a dedicated room for acquisition, preliminary and final read is recommended (Fig. 6.12). A dedicated CBCT room has the following advantages:

- Possibility of lowering light conditions for a better study review
- Dedicated monitors (usually larger, better oriented and in some cases specifically built for medical radiographing)
- Computers with better processors (lower rendering time)
- Privacy

Whenever possible, reviewing the whole CBCT study with the patient is recommended as it helps with explaining the findings, diagnosis, treatment complexities, and different alternatives.

Case Presentation

Once all the tests have been performed, data gathered and a conclusion made, the patient is brought back to the seated position. The patient chair is rotated so the patient and dentist face each other. At this moment, the case will be presented.

In order to present the case, the operator uses all the information collected during the consultation, which has been contemporaneously introduced into the patient's chart. A PC monitor placed within the circle of influence of both the operator and the patient (they are both facing directly this monitor) is used to show all the pictures and radiographs taken. Intraoral pictures and radiographs are useful tools to illustrate the actual problem in a didactic manner. A drawing option/program is also helpful to point out important aspects of the case or the treatment, or to annotate any images taken. The diagnosis, prognosis, treatment, alternatives, and fees are discussed at this point. Any recommendations to the patient are also given. The assistant enters details of the conversation in the patient chart whilst the discussion is taking place. Time has to be allocated for the patient to ask questions and to explain any concerns about the situation. In a professional environment, the patient leaves the office having a good understanding of the nature of the problem, the different treatment options, and the expected fees. The patient should have confidence in the entire experience, ranging from the behaviour of the office staff, the nature of the treatment involved and the operator's chairside manner.

After the Appointment

When the patient leaves the operatory with the assistant and goes back to reception, all relevant information should be quickly sent to the receptionist. With the right type of practice management software, this can be completed before the patient leaves the operatory. This is so details about rebooking the next appointment can be correctly entered (e.g. what type and duration of appointment is required). Accurate fee processing and payment can also be taken without delaying the patient. In some countries, payments are not taken at the office and a bill is sent to the patient. However, in most situations, payment is taken on the day of the appointment before the patient leaves, if not already pre-paid. If there are details requiring an external provider (e.g. the patient needs an appointment to see their referring dentist), this can be reinforced to the patient—or even assisted by calling the referring office on the patient's behalf.

Clinical Microscope Endodontics: Non-surgical Treatment

7

Abstract

This chapter focuses on the component of endodontics that takes a majority of clinical time – non surgical treatment. The path to your team combining a set of individual movements into a procedural sequence is discussed. Focusing on minimising wasted time and movement is the foundation of efficiency; strategies for your team to improve efficiency are covered.

Begin with the end in mind—proverb

Key Points

- A treatment procedure is a sequence of individual steps and single exchanges.
- Being prepared for adaptation or a protocol deviation is how to strengthen the skillset, by storing new patterns to be recognised.
- Wasted movement and time hinders or prevents the procedure from occurring; team members should be aware of all surroundings for each step.

Introduction

Non-surgical endodontic treatment forms the significant majority of your professional career spent chairside. Therefore, this component should have the most focus in order to maximise teamwork and efficiency. Because the procedure generally follows an identical sequence, consistent patterns can be stored which help operators and assistants develop expertise. When each exchange is broken down into individual movements, readers will find that there are a limited type and number of instrument and material passes. A treatment procedure is just a sum of single exchanges in sequence. Many different instruments will use the same motion to pass and receive, even though their use may vary. By keeping the overall set of

K. E. Yapp, C. Aznar Portoles, *Team Microscope Endodontics*,
https://doi.org/10.1007/978-3-031-70188-7_7

movements simple, this helps new users master the procedure. Once the main sequence of events is learned and stored into patterns, expertise is developed by recognising when changes occur, adapting to these changes and incorporating these changes into the core pattern. The operator should be aware of what the assistant is doing and vice versa. Situational awareness is a characteristic of expert performance and experts will be able to recognise the stage of the procedure, respond without delay, and quickly make any adaptations as required.

Before the Appointment

When the operator is planning the clinical case, relevant parts of the treatment procedure should be identified, so the specific nuances can be adequately addressed. For example, if there is subgingival interproximal caries on a molar, caries removal plus soft tissue management will need to be done, plus managing the matrix situation for the restorative material. Before treatment is started, the general overall sequence should be planned in your mind, taking into account all information—including CBCT imaging. Anticipating canal location and relative angles of anatomical structures should be a part of the overall planning process. To communicate this to your team, it should be written down if this is an unfamiliar procedure for either team member. This ensures the intent is clearly communicated and all protocol deviations and specific case components are accounted for. A written list also helps a new or less experienced assistant by providing a reference, in the event a step is forgotten.

The instrument and material setup should be completed before the patient enters the operatory. By ensuring all required instruments are present and ready, it avoids any unnecessary delay in the procedure of the assistant having to stop, gather what is required—or leave the room—to find it. Wasted time is a wasted movement. There may be some time-sensitive materials that set quickly, so being able to properly handle such materials before they set, will also ensure patient outcomes are not compromised.

Setup should be considered in separate "zones" so that each section can be prepared for and completed before moving onto the next. These include the: back wall tray, dental cart, locked patient chair position, monitors (ensuring the correct patient file is open in the practice management software), and operator/assistant stool positions. This is so the patient can enter the operatory, sit and talk with the assistant, then the operator can enter and greet the patient without having to rearrange any equipment or furniture.

Ambient settings, such as lighting, music, and temperature, should also be adjusted to ensure a comfortable environment.

Start Treatment Checklist
Sample cart setup
High speed handpiece—round diamond
High speed handpiece—tapered round ended fissure diamond
Slow speed handpiece—lentulo spiral
Ultrasonic tip—angled diamond tip
Ultrasonic tip—90 degree fine tip
Stropko tip—air/water
Stropko tip—air only
Sample back wall setup
Treatment cassette—instruments out and on bench
Matrix band to suit tooth, wedge (check pre-op PA)
File pack (new #10 and #15K files) with sponge and ring
NaOCl with needle bent
Suction tips: green, then white with Endo-Eze tip attachment
XDR sensor with ComfortWand holder
RD with clamp and frame
LA syringe setup as per tooth, with topical LA on cotton bud
Rotary handpiece
Apex locator cords
Cavit, corner of file sponge removed
Ultracal
Temporary material (Fuji IX, Fuji II LC or flowable composite)
Articulating paper
Gauze
Floss

Complete Treatment Checklist
Sample cart setup
High speed handpiece—round diamond
High speed handpiece—tapered round ended fissure diamond
Slow speed handpiece—lentulo spiral
Ultrasonic tip—angled diamond tip
Ultrasonic tip—90 degree fine tip
Stropko tip—air/water
Stropko tip—air only
Sample back wall setup
Treatment cassette—instruments out and on bench
Matrix band to suit tooth, wedge (check pre-op PA)
Patient file pack with sponge and ring
NaOCl with needle bent

EDTA with needle bent
Suction tips: green, then white with Endo-Eze tip attachment
XDR sensor with ComfortWand holder and Rinn post-op PA holder
RD with clamp and frame
LA syringe setup as per tooth, with topical LA on cotton bud
Sealer
GP cones in measuring block
Paper points—2x per canal
System B tip and Elements cartridge
Etch with microbrush
Restorative material setup: If Luxacore = Centrix tip and syringe, photo bond and microbrush, dry microbrush, fibre post
If Amalgam = amalgam, instruments, damp cotton bud
Articulating paper
Gauze
Floss

During the Appointment

The ideal team strategy is to be aware of and anticipate the entire procedure from start to finish, but to focus and communicate about the current step and only a few steps ahead. Prioritising correctly will follow this thought process because the logical explanation is that each step must be completed before the next step begins, i.e. the assistant would not be mixing amalgam, or focusing on packing up all the instruments during local anaesthetic administration.

Another method of being efficient is to make all instrument passes into exchanges where possible. This means the assistant has the next instrument or material ready to exchange with the operator. Anything else would be a lot of wasted movement, i.e. the assistant should not be taking one instrument, putting it down, getting the next instrument and passing to the operator. The responsibility also lies with the operator, who has to clearly communicate what is required (if a deviation from protocol) to the assistant before returning the currently used instrument.

Whilst the individual steps and general training procedural sequence have all been covered in detail in Chap. 5, this will go through the entire clinical procedure of non-surgical endodontic treatment, divided into two parts: start and complete.

Start Treatment

The beginning of the procedure should always begin with warmly greeting the patient and making them feel at ease. By facing the patient whilst entering the operatory, this gives a confident approach and does not scare or startle the patient.

Greeting the patient from behind will never provide a good impression; no one enjoys having someone creeping up behind them, not even from their dentist. This initial time for communication should not be rushed—briefly explaining what will occur during the appointment is critical for building trust. Even if this has been covered at the consultation, reinforcing these details will only help with communication. If not done already, the assistant should place the protective patient bib on and provide the protective sunglasses (or exchange with the patient's glasses) before the patient is reclined into the preset position. If this is not done and the operator is ready to start when the patient is reclined, this prevents the team from performing the next step of the procedure.

With the patient comfortably reclined, the operator brings the microscope and dental cart into position. It should be emphasised that having the microscope in the right position, so that the operator's eyes are slightly back from the eyepieces, is important. This allows the operator to move their eyes in and out of the microscope as needed, without having to move from the hip (Fig. 7.1).

The large mirror is passed to the operator—the same as the final step of a mirror change. At the same time, a class 3 movement to take the Stropko syringe is done, to dry the mucosa. This prevents the saliva from washing out the topical anaesthetic. The Stropko is returned to the cart and local anaesthetic administration is performed, starting with the cotton bud with topical anaesthetic passed like an instrument pass. The assistant immediately takes the LA syringe in the left hand. When the cotton bud is returned like a file pass (pointing towards the assistant's left shoulder), the

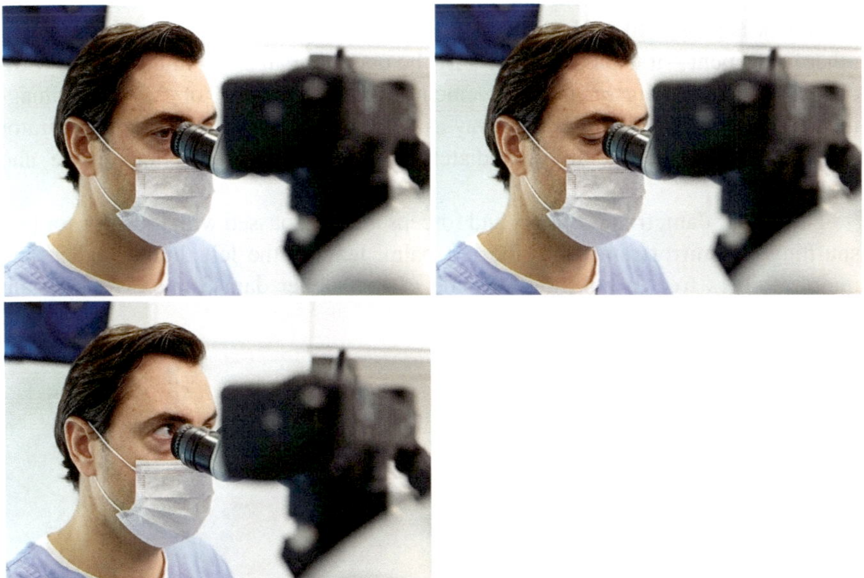

Fig. 7.1 (i) Operator looking into the microscope with eyes slightly back from the eyepieces. (ii) To look towards the patient, a downward eye movement is done. (iii) To look up and across, an upwards sideway eye movement is done

assistant takes it with the right hand before the operator's signal to receive is done. The operator's right palm is now facing up and ready to accept the syringe. The assistant passes it with the left hand and once the syringe is safely exchanged, the cotton bud can be discarded. It is important to pass the syringe before throwing away the cotton bud, not the other way round; otherwise, the operator will be waiting and the procedure is delayed. During LA administration, the assistant can see in the binoculars if saliva is pooling and should proactively suction without being asked. Entering data about the LA used into the patient record should be done at this time, which does not require more than a few mouse clicks. Because the assistant side monitor and mouse are in the circle of influence, this can be done without diverting attention away from the operating field and the patient. The first LA carpule is exchanged, the second is administered and once complete, the Stropko is used to wash the patient's mouth with the assistant suctioning.

Local anaesthetic video link:

After the local anaesthetic has had time to work and the patient is numb, the smaller (working) mirror and bite block are passed to the operator (Fig. 7.2). The bite block is passed like a hand instrument following the same principles as every other instrument—it is passed in the way the operator can use it immediately, without any additional adjustment or movement. Because it is a "U" or "V" shape, imagine that being oriented in the same way as the patient's jaws. That way the operator can place it in the mouth immediately without wasting additional time and movement.

The rubber dam, frame, clamp, and forceps are then passed with the operator first shuffling the mirror down into the left palm, leaving the left thumb, index, and middle fingers free to hold the left side of the rubber dam frame. The assistant

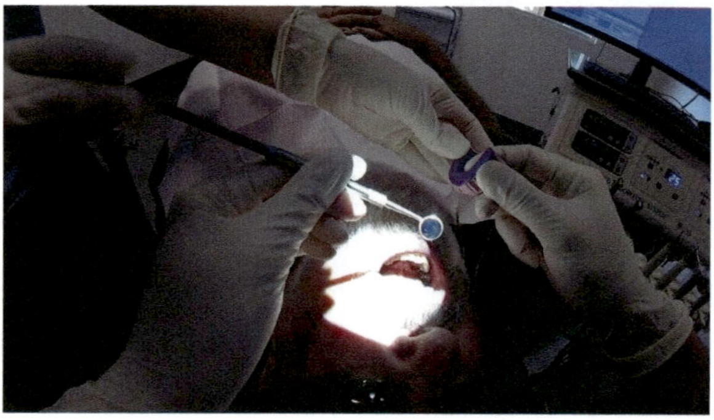

Fig. 7.2 Bite block passed similar to an instrument pass

should always pass this in the same way for all teeth, which is to align the frame straight with the patient's mouth. This is because the frame will always be in the same position, regardless of the tooth being treated. The operator's left hand will hold the left side of the frame first, then the right hand will move to hold the forceps. When this is roughly in place, the left hand fingers can release the frame, shuffle the mirror up to the working position and into the operating field, where the operator's eyes then move to look in the microscope. Here, rubber dam clamp placement is guided with precision by looking in the mirror, ensuring that the tines of the clamp are securely on tooth structure. Once secure, the forceps are released and the fingers of both hands used to "rock" the clamp, making sure it does not pivot or slip. The forceps are returned to the assistant with a palm up motion, which will then be the same motion in reverse for the assistant to return the forceps at the end of the appointment. The dam is stretched over the frame and if isolating more than one tooth, the dam is stretched and contacts flossed as required. Once the operating field is ready for treatment, the *starting protocol* is performed (see Chap. 6).

Treatment starts with the high speed handpiece either: removing restorative material, accessing through a crown, or accessing an intact tooth (such as a traumatised incisor). The two methods of the assistant managing the field are with close suctioning and mirror exchanges, or an assistant air/water Stropko syringe. Both methods should be learned because both techniques are valuable in situations where the other cannot achieve a desired result. Both scenarios also require the mirror to be placed further away from the tooth. This is another advantage of a wide field of view and low magnification—the mirror can be held further away, limiting the amount of spray on the tooth. It should be noted that this part of the treatment is the same as at the end, when the restoration is being polished and occlusal adjustments made.

The mirror change situation has the water turned on at the handpiece, with the level of water spray being low to moderate, where there is just enough aerosol to clear the bur of debris during cutting. The operator holds the mirror with the handle close to upright, so there is room for the assistant to place the suction underneath. The suction is held in the assistant's left hand, with the new mirror *being held in the exact position it is to be placed* in the right hand. Most importantly, both arms should be on the armrests for both operator and assistant. This next step is crucial because it determines how quickly the mirror gets dirty, how well an assistant is focusing on the task at hand, and how much time and movement is wasted throughout the procedure. It relates to *precision suction placement* and procedural efficiency. This is the most common error or factor that leads to slow and inefficient high speed handpiece use. For a majority of dentists, they will start by cutting with a diamond bur with water spray—which dirties the mirror to the point of being unusable within a few seconds. This prevents the operator from seeing, prevents the team from completing the current task, and stops the procedure. By placing the suction right next to the tooth, it catches most of the water spray before it has the opportunity to dirty the mirror. This can only be done with the assistant looking through the assistant binoculars. When the mirror does get dirty *and* there is a break in usage of the handpiece or ultrasonic, then the mirror change occurs. Because the assistant

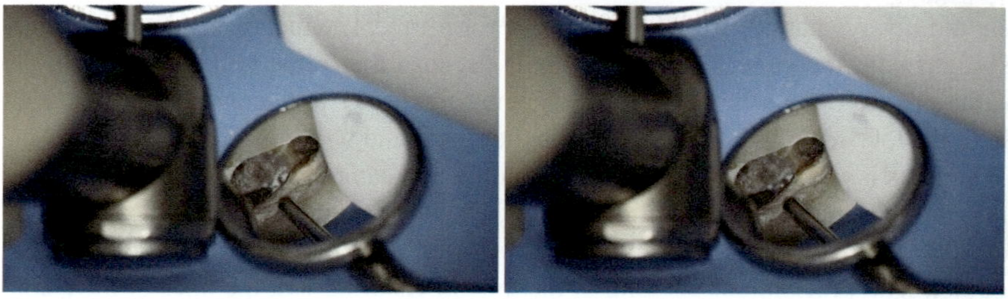

Fig. 7.3 Before and after air-only Stropko has cleared dentine dust, providing complete visualisation of the pulpal floor

sees exactly what the operator does in the microscope, the mirror change is proactively directed by the assistant.

Using the assistant Stropko has two main goals: 1) to clear the bur and mirror using a precisely placed water stream and 2) to blow dentine dust away with a focused blast of air (Fig. 7.3).

These ensure no interruption of the procedure in both settings and complete continuous visualisation. This cannot be achieved with a standard triplex air/water syringe because the tip is neither narrow nor precise enough to deliver air or water in the right location. The suction is held in the assistant's *right* hand because there is greater reach by holding the suction handle lower down. The Stropko irrigator is held in the assistant's *left* hand because it is closer to the operating field. If the Stropko was in the right hand, in most cases it would be too far away from the mouth when the arm is on the armrest. It is activated with the index finger to ensure a relaxed position and controlled placement. Pressing the button with the thumb has less control. Most importantly, assistant positioning must ensure both arms are on the armrests. The assistant seated position will likely have to move closer to 3 o'clock, so that the right arm is close enough to the patient's mouth. Separate air/water and air-only Stropko syringes must be used. After water has been delivered in an air/water Stropko irrigator, water becomes stuck in the Luer lock tip. When air is then expressed, it creates mud when mixed with the dentine dust and hinders visualisation. The first part of the procedure (bulk restorative material and/or dentine removal) uses water and air for the mirror. When finer exploration is required for canal location, the assistant switches to the air-only Stropko irrigator.

Once the canals are located, hand files and sodium hypochlorite irrigation are used for canal negotiation. In some instances, rotary files would be used for coronal preparation before working length is determined. This initial canal "instrumentation" phase involves changing from hand files, to irrigants, back and forth in varying sequences multiple times. Whilst the exact sequence depends on the tooth, it is important for the assistant to have the ring with file sponge ready on the left index finger, holding the irrigation syringe in the left hand and prepared for an immediate exchange from hand file to irrigation (and vice versa), all in one motion. When the canals are being irrigated, the assistant uses the bendable Luer lock tip placed on the white fine suction tip inside the pulp chamber. This suctions the irrigant and prevents an aerosol from being created. The aerosol will dirty the mirror and lead to inefficiency. It is important for the operator to always use the index finger to aspirate the plunger slightly prior to returning it to the assistant with a 1–2 motion, preventing any drops of irrigant falling onto the patient. The assistant should check for any air in the irrigation syringe and remove the air, by holding the needle up and lightly tapping the plunger onto the back wall bench. If this is a retreatment case, chloroform would be used in a syringe and all passing motions between syringe and hand files would be the same.

In some situations, particularly lower molars, there would be a tendency to try and hand file the canal by raising the elbow and shoulder. This is colloquially termed the "chicken wing" and is due to the angle of the canal directed upwards and the limitation of the wrist to accommodate this angle. The way to prevent this is simple. The operator turns the patient head (e.g. to the right for a lower right molar) and compensates by swivelling the patient chair to the left. This keeps the tooth in the field of view and does not move the microscope. For a lower left molar, hand filing with the left hand, once the file is placed in the canal, can be done to overcome the risk of the chicken wing position. If any adjustments to the patient position need to occur, slight head movement and swivelling the patient chair in the opposite direction can be done.

Once patency has been estimated and working length is to be determined using the apex locator, often a 25 mm (long) #10K file is used. This gives enough room for the rubber stopper to be moved. After this is placed in a dry canal, the assistant will pass the lip clip to the operator's right hand, to be placed in the corner of the lip. If it is placed on the patient's left side, it can be exchanged into the operator's left hand after shuffling the mirror down into the palm. The important step in using the apex locator efficiently as a team is to move the patient's head and use direct vision—do not move the microscope. For a molar, the occlusal table and cusp tips should be directly visualised and sometimes the patient head should be angled back slightly to ensure all of these landmarks can be seen. The apex locator "fork" attachment is passed from the assistant's right hand to the operator's left hand, like a hand instrument. The left-handed pen grip is the most relaxed way to activate the apex locator and control the rubber stopper at the same time. The operator's right hand is moving the hand file either with the thumb and index finger, or with a file holder (k1forceps.com). The file holder allows greater control, precision, and visual access when space is limited. When using the apex locator, the operator should be looking at the file and rubber stopper. If there is a metal crown, the file should be controlled so that it does not touch the crown on the side of the access cavity. By focusing with the microscope, this can be easily prevented. The assistant is looking at the apex locator screen and reading out the number value as required. Some units may have a sound which helps identify the positional value; however, the key part of this step is that the operator is not turning the head to look at the apex locator screen. The lengths should be measured with the ruler handle in the microscope (Fig. 7.4). If the ruler is out of focus, this would be a depth of field issue determined by the objective lens focal length and/or dual iris setting (see Chap. 3).

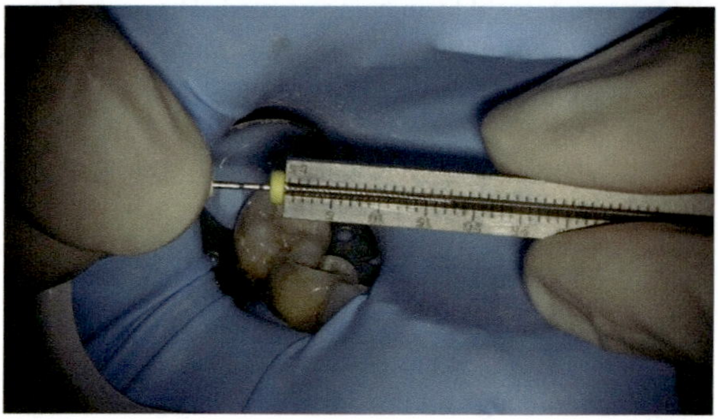

Fig. 7.4 Measuring file length using the microscope

Once the estimated working length values are entered in the patient record, the files are placed in the canals—passed in their normal way—and then the working length radiograph is taken. The bite block is removed and along with the mirror, returned to the assistant. The plastic rubber dam frame allows the frame to be kept on without superimposing onto the files. For lower molars, sometimes the opposite corner of the dam has to be loosened, to reduce the tension on the dam. This is because the tongue can push against the X-ray sensor or phosphor plate, into the rubber dam clamp and tear the dam. The clamp position is a good reference for where the X-ray tube and sensor should be positioned. A clear understanding of the spatial relationship between X-ray tube, tooth, and sensor should always be obtained, prior to taking the radiograph, to ensure the tooth is not missed. Once this is taken, with the patient holding the sensor handle, and the microscope and patient chair are returned into position, the mirror and bite block are passed from the assistant to the operator, just like at the start of the appointment. Consistency of passing movements will only strengthen pattern recognition for team members.

Taking radiographs video link:

The working length values are now confirmed, entered into the patient file, and rotary preparation can occur. After removing and returning the files from the working length radiograph, the assistant should pass the irrigation syringe and the pulp chamber then filled with hypochlorite. A cordless rotary handpiece is the ideal type for rotary files, which should be held by the assistant on the back end, away from the file. This prevents any accidental starting of the handpiece during passing. When the assistant loads the rotary file, the rubber stopper should be pulled slightly away from the shank because it is easier to adjust the stopper using the mirror handle when there is distance between the stopper and the shank. The operator holds the cordless handpiece like a hand instrument. Because there is no serrated grip to hold onto, the assistant should aim to have the start/stop button positioned in line with the operator's index finger (for maxillary teeth) or thumb (for mandibular teeth). The difference in passing is similar to the irrigation syringe—either the rotary file is pointing up or pointing down, depending on the tooth, and all other factors are the same. If the rubber stopper landmark needs to be visualised using the mirror, the angle can be slightly changed so there is clear line of sight to the rubber stopper and the respective cusp tip. When the rotary file needs to be cleared of debris, the motion is similar to clearing a hand file of debris—it is pointed towards the assistant's left shoulder and held still. The assistant then pierces the file with the sponge and withdraws. Usually, one pass is enough to clear all the dentine debris from the rotary file. Once the rotary file has been used and the canals need to be irrigated, the assistant should have the irrigation syringe ready in the left hand, before taking the cordless handpiece in the right hand. The operator returns the rotary handpiece with a 1–2 motion, just like a hand instrument. This sequence can then be repeated, with the assistant changing rotary files, as many times as required. Once the rotary canal preparation is complete and no further canal location or exploration is required, the intracanal medicament is ready to be placed. Aspirating all canals after the final irrigation is done, dries the canals effectively. The syringe is returned with a 1–2 motion and the chamber dried with the air from the Stropko syringe. If a lentulo spiral in the slow speed handpiece is used to deliver the medicament, taking this from the cart is done using a class 3 motion. The assistant would then dispense the medicament onto a mixing pad, hold it in the field of view so the operator can collect it on the spiral when looking in the microscope, then deliver into the canal. If a delivery tip syringe is used, the assistant passes this just like an irrigation or flowable composite syringe.

Temporisation can be simple or complex depending on the restorative situation. Simple situations require no matrix band to build up a proximal wall. In this scenario, Cavit is placed on the tip of a flat plastic instrument, in the same angle as the blade. The width of the Cavit cylinder is slightly narrower than the diameter of the access cavity. The assistant passes this hand instrument and the operator places the Cavit immediately. The assistant knows how wide to shape the Cavit because of looking through the assistant binoculars. A small corner of the file sponge is removed and moistened to clean the Cavit around the edges of the access cavity. This is passed in locking tweezers, just like a hand instrument. After returning it, the tooth is dried with the Stropko irrigator and ready for the top layer of temporary restorative material—likely flowable composite or glass ionomer cement (e.g. Fuji IX). If flowable composite is used, this is delivered and then adapted with a microbrush that the assistant has ready immediately after passing the composite syringe. If glass ionomer cement is used, this is delivered, then adapted with a slightly damp cotton bud.

When rebuilding a proximal wall, the Cavit is placed first as above. The matrix band retainer (i.e. Tofflemire) is passed like a hand instrument. It is placed with direct vision in the microscope by slightly changing the patient head angle. If a corner of the matrix band needs to be held away from the tooth so the band can seat, the assistant passes the explorer which is used to stabilise the band before the left hand is used to seat the band (Fig. 7.5).

After tightening the retainer, adaptation at the floor of the proximal box is checked with the explorer. If there is a gap, then a wedge is placed. The assistant holds this in locking tweezers before passing like a hand instrument. If it is a triangular shaped wedge, the base should be placed on the side closest to the gingiva. Depending on the situation, sometimes the wedge may need to be loaded into the locking tweezers backwards (wedge pointing towards back end of tweezers), so that it can be placed easily from the buccal or lingual side. All of these movements occur by looking with the mirror in the microscope. Proximal box adaptation to the matrix band is checked again using the explorer. If a different type of matrix system is used, the components can likely be placed with locking tweezers—all of which are exchanged like hand instruments. Matrix band placement in this situation will be the same as for the complete appointment.

After the material is placed and set, wedge, and band removal is simply the opposite of placement. Polishing the occlusal surface is an identical process to using the handpiece at the start of the appointment. Removing the rubber dam is the opposite of placement, where the forceps are in the same position for the operator when they were initially given back to the assistant. Once the bite block is removed, the patient should have an opportunity to swallow and rest their jaw. The assistant should then have both the large suction (in the right hand) and articulating paper forceps (in the

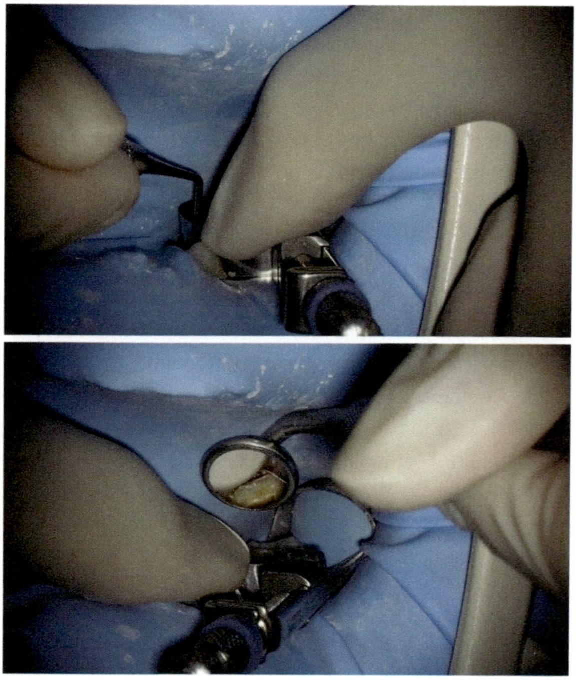

Fig. 7.5 (i) Explorer helping matrix band seat around tooth margin. (ii) Checking band is properly seated before tightening matrix band

left hand) ready. The operator should get the Stropko syringe with a class 3 motion, dry the tooth with the suction next to the tooth, and leave it there when the operator places the Stropko syringe back on the cart and receives the articulating paper from the assistant's left hand. This ensures the tooth does not get wet and the articulating paper markings will be transferred. If any occlusal adjustment is required, the process is repeated and all of this is done with both operator and assistant looking through the microscope. During the entire appointment, the only real movement from the operator is when taking the radiograph. Everything is done in the microscope, efficiently and without wasted time or movement.

Complete Treatment

The beginning of this appointment has many similarities to the start treatment appointment. The main difference is that the operator should ensure that symptoms have resolved, by checking percussion with the back of the periodontal probe. The assistant passes this like a hand instrument, but backwards. If there is no percussion tenderness, the treatment is ready to be completed.

After local anaesthetic administration, rubber dam and restoration removal, irrigation in the canals to remove the medicament occurs. To ensure the canal is free from all debris, this is done twice, the first time with #15K hand files, with irrigation immediately after, then using the master apical rotary file *by hand*. Each passing movement with the assistant should be with the irrigation syringe in the left hand for an immediate exchange—no different to the start treatment appointment. Measurement of each file is checked with the ruler handle whilst looking in the microscope. Once the irrigation is complete and the canals are free of debris, fitting the gutta percha (GP) cones is done next. A general rule should be that the order of canals being fitted or filled should be the same for each tooth, i.e. an upper maxillary molar should be in the order of: MB1, MB2, DB, and then P. A mandibular molar should be: MB, ML, DB, and DL. These are examples to show that the same order should be done every time for consistency and predictability. The assistant should use a measuring block and pick up each GP cone with locking tweezers (Fig. 7.6).

This is passed to the operator like a hand file, who confirms the length on the ruler and squeezes the cone with a bit more pressure leaving a "crimp" mark. This

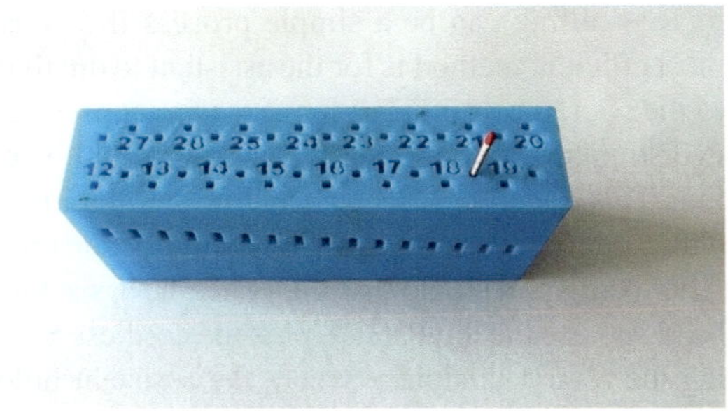

Fig. 7.6 GP cone in measuring block

is so there is a clear visual reference of the length in relation to the reference point. After passing each GP cone, the assistant immediately takes the next GP cone, ready to take the old (empty) tweezers with the left pinky and pass the next tweezers with the left hand—all in one motion. When this is repeated with all GP cones, the check radiograph is taken. The process of the working radiograph is the same as in the start treatment appointment, with the bite block removed and mirror returned to the assistant first.

If the GP cones are at the right length, no adjustment is required. If the cone is too long, the tip would need to be cut with scissors. This occurs by the operator taking the GP cone with the locking tweezers above (coronal to) the crimp mark, so the mark is not modified. The tweezers are then passed to the operator's left hand—with mirror shuffled down into the palm. The assistant passes the scissors into the operator's right hand, with thumb and ring fingers in the two rings. The operator's middle and index fingers are supporting the side of the scissors for stability. After the GP cone is cut and the scissors returned, the GP cone is returned to the right hand and tried in the canal again, to see if the cone is seating to the same length as referenced by the crimp mark. This is another reason why not to modify or displace the marking on the GP cone.

Final irrigation involves EDTA, followed by hypochlorite. This is usually the only time one irrigant is exchanged for another. After the assistant passes the EDTA syringe, the hypochlorite syringe is held in the left hand. When the operator returns the EDTA syringe with the 1–2 motion, the assistant takes it with the pinky and immediately passes the hypochlorite syringe in one motion. When all irrigant is delivered, all canals are aspirated and the assistant places the Luer lock tip over each canal orifice, so the pulp chamber and majority of the canals are dry. The operator takes the Stropko with a class 3 motion and dries the chamber.

Paper points in the locking tweezers are used next, and these are just a sequence of hand instrument exchanges. The assistant should be aware of the working lengths and hold the paper point at a length slightly longer than the confirmed length. If there is an 18 mm canal, the paper point should not be held at the very end because there likely will be insufficient room in the mouth. The back end of the paper point will hit the opposing teeth, the tip will hit the adjacent teeth and the paper point will be bent and unusable. These types of overlooked details can lead to a lot of wasted time. Once the canal is dry, if a post is going to be cemented, this should be fitted in the canal at this point. The post space length will then be measured, and this will determine how much GP to remove for the post.

Root filling can be a simple process if a consistent sequence is followed. The most efficient method is for the assistant to dip the GP cones in sealer before passing to the operator. If preparation sizes are conservative, meaning GP cones have limited strength, applying the sealer first should be done instead to avoid cones buckling. This is similar to using a lentulo spiral for placing intracanal medicament. With either option, each canal stage should occur one at a time, using the same order. After being passed the first GP cone, the assistant will take the tweezers in the right hand and use the left hand to pass the cordless System B unit. This is similar to passing the rotary handpiece where the assistant holds it at the back end and aims to place it in the operator's hand with the button in line with the thumb or index finger. The operator burns the GP cone at the orifice, so that there is good visualisation into the pulp chamber. The System B unit is exchanged with the next GP cone. The assistant uses a similar method by taking the System B in the right hand and using the left hand to pass the GP cone. This is repeated until all canals have had the GP burned at the canal orifice.

Downpacking with the System B is now done, where the assistant has the Buchanan plugger ready. Similar to the previous set of exchanges, after the operator has downpacked the GP to the required level, the assistant immediately passes the Buchanan plugger to the operator, who applies pressure to the GP. If a post is being placed in the canal, the measured post space length is used as a reference for how deep the downpack should be. By having this predetermined length, this prevents time and movement being wasted by downpacking, checking the post fit, repeating and not having the post fit to an ideal length. The downpack is repeated for all canals until the next stage—backfill. The cordless Elements unit is ideal for extruding warm GP and the passing sequence is the same for the System B unit. One canal is filled at a time, with the Buchanan plugger exchanged so that immediate pressure on the GP can be placed. Often, multiple small increments of GP will need to be placed and compacted between each one. If the canal angle requires the Elements tip to be bent (likely MB roots of upper molars and mesial roots of lower molars), tweezers are passed from the assistant's right hand to the operator's left hand with beaks facing the operator. The left hand bends the Elements tip and passes it back to the assistant with a 1–2 motion. After the canals have been filled, a check radiograph is taken prior to placing the final restorative material.

The restorative scenario is similar to the start appointment, except the material is different. If a matrix band is needed, this is placed first before cleaning the pulp chamber with etch and scrubbing with a microbrush. If restoring through a crown or other situation where no matrix band is needed, the same steps apply but without using a matrix band.

Amalgam is placed with the assistant placing the amalgam from the carrier into the tooth. This is the most efficient way to place it as a team and also to avoid the amalgam setting before it can be placed in the tooth. To ensure the first two increments are properly adapted to the pulp chamber floor, the Buchanan plugger is used first. The sequence is: assistant places amalgam whilst looking in the microscope, operator packs it using the Buchanan plugger. This is repeated once, then the thinner Buchanan plugger is exchanged with the regular amalgam plugger—just as a hand instrument exchange. Amalgam is then packed with the same sequence until filled. If a matrix band is used, the wedge(s) are removed first, the Tofflemire retainer is loosened completely from the band, then the assistant passes the larger end of the amalgam plugger. The operator uses this to hold the marginal ridge whilst carefully sliding the band out from the contact, preventing fracture of the marginal ridge. The assistant then exchanges the hand instruments with a Hollenback carver, ball burnisher, and finally a damp cotton bud. In some situations, the operator will need to cut the interproximal amalgam back to ensure it is flush with tooth structure. This can only be done with the assistant Stropko clearing the debris from the bur. By using an air/water spray from the handpiece, proper visualisation of the deep margin is not possible. Once the carving is finished, the rubber dam is removed and the occlusion checked. After any adjustments are made, the post-operative PA radiograph is taken.

Composite is typically placed with a Centrix syringe, with or without a fibre post. After washing away the etch, the bonding system is applied and cured. The microscope orange light filter to inhibit setting should be changed. The microbrush is passed like a hand instrument and the bond brought into the operating field so the operator can dip the brush into the bond, all whilst looking in the microscope. After curing by the assistant, the Centrix tip should be passed like a hand instrument, the tip bent by the operator using the mirror handle ruler, then tested in the canal to ensure the tip is fitting and deep enough to touch the GP. If a fibre post is used, this should be checked again, with the correct length measured with the locking tweezers. The assistant should have the following ready because of the quick setting time: dual cure composite, fibre post and dry microbrush. As soon as the Centrix tip is loaded with the composite and passed to the operator in the composite applicator, the fibre post in the locking tweezers should immediately be ready in the assistant's left hand. After this exchange, the microbrush is ready for the operator to adapt the setting composite before light curing. Once cured and matrix band removed, the restoration is polished and occlusion checked before taking the post-operative PA radiograph.

After the Appointment

Once the radiographs are taken and the treatment is complete, the patient chair is returned to the preset upright position, post-operative instructions are given and the bib and glasses are removed. It is important to emphasise that some soreness is to be anticipated for a few days. The severity and duration of post-operative pain is also usually dependent on the amount of pre-operative pain, which should help guide patient expectations. The main feature for both team members is that a majority of the entire treatment requires minimal movement and most importantly, no wasted movement. Because everything is set up and prepared, delays do not need to occur. Positioning of the back wall and dental cart to have everything in the circle of influence is crucial for treatment to be efficient.

Clinical Microscope Endodontics: Surgical Treatment

8

Ivan N. Vyuchnov

Abstract

Endodontic microsurgery requires a different way of thinking because of the different biological factors and treatment goals. This chapter discusses the need for a refocused team approach, including working with two assistants. The importance of dynamic patient positioning to ensure efficiency is covered. Specialised instruments with design features required for microsurgical treatment are described.

Fast is slow, but continuously, without interruptions—Japanese proverb

Key Points

- Two assistants and an operator are required to ensure efficient and precise endodontic microsurgery is performed.
- Dynamic patient head movements for the different procedural stages give the operator direct vision, without wasting time and movement.
- Crypt isolation by the first assistant looking through assistant binoculars provides adequate visualisation and a dry field for root end material placement.

The original version of the chapter has been revised. A correction to this chapter can be found at https://doi.org/10.1007/978-3-031-70188-7_9

I. N. Vyuchnov (✉)
Moscow, Russia

K. E. Yapp, C. Aznar Portoles, *Team Microscope Endodontics*, https://doi.org/10.1007/978-3-031-70188-7_8

Introduction

Team endodontic microsurgery has different procedural goals to non-surgical treatment. Although the aim of achieving a favourable patient-centred outcome is the same—to preserve a tooth without any functional impairment, where the parameters are those valued by the patient—the entire mindset with a microsurgical approach requires a refocused treatment strategy for the team.

Because extended treatment time and associated factors (an open surgical wound, prolonged bleeding) are critical in affecting healing outcomes and morbidity, the microsurgical team needs to be even more efficient than the non-surgical team in order to provide the same level of care. This means two chairside assistants are needed, each with different roles (Fig. 8.1). The operatory setup also requires some changes to ensure both assistants are able to perform at the highest level of teamwork and efficiency. The instruments and materials are also different, but similarly to non-surgical treatment, are arranged in the sequence of use to ensure time is not wasted and the procedure occurs in a logical order.

Minimal muscle movement is important for the operator and both assistants because it helps reduce stress during treatment, alleviate tremor, and increase concentration. Muscular "at rest" positions are facilitated by microsurgical stools with armrests, just like in non-surgical treatment. Physiological tremor is affected by gravity and proper support is required to help stabilise the forearm. These are essential for all team members in being able to perform microsurgery at the highest level.

Visual acuity, the ability of the eye to distinguish object details at a given distance and is significantly improved with the microscope. According to Shanelec (2003), the resolving power of the unaided human eye is only 200 μm (0.2 mm). At this level of visual acuity, the greatest possible accuracy for human hand movement is 1000 microns (1 mm). Physiologic tremor can further reduce the accuracy of movement to 2 mm. At 20× magnification with the microscope, the accuracy of hand movement approaches 10 μm and visual resolution approaches 1 μm. Therefore, greater precision in human motor control is achieved using the microscope.

Precision is a critical component in microsurgery because the amount of tissue damage corresponds to the healing outcome and post-operative morbidity. An ergonomic team microsurgical approach provides a better environment for achieving

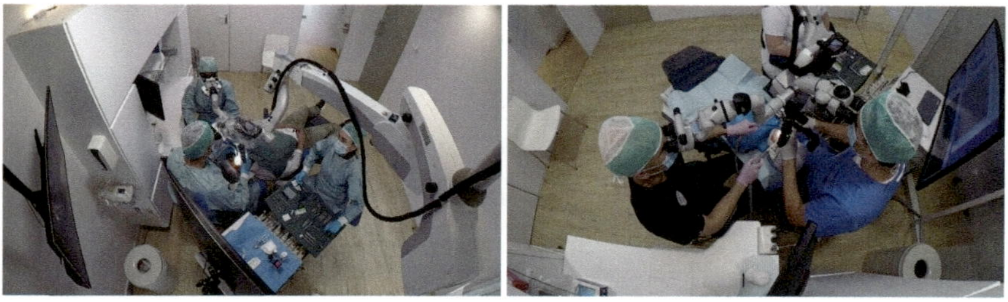

Fig. 8.1 Endodontic microsurgery team with two assistants. Note all team members have both arms supported on armrests

ideal microsurgical goals. These are: minimally invasive and correctly placed incisions, atraumatic flap reflection, proper visualisation enabling correct root tip resection (90 degrees to the long axis), a detailed inspection of the resected root surface anatomy (canal isthmus and accessory anatomy), canal preparation along its long axis using ultrasonic tips, precise placement of the root end filling material and accurate soft tissue management by placing microsurgical sutures. There have been advantages to a microsurgical approach regarding soft tissue healing in a periodontal setting. Microsurgical connective tissue grafts were found to have significantly greater vascularisation and recession coverage, when compared to those performed with a macrosurgical method (Burkhardt and Lang 2005). Microsurgical wounds had significantly lower levels of serum interleukin inflammatory markers compared to conventional scalpel wounds made without a microscope in an animal study (Hirata 2012). The low levels of inflammatory markers are also achieved by the first assistant rehydrating the flap using a Stropko irrigator. The microsurgical approach has clear advantages to the patient.

The first surgical assistant is primarily tasked with field management by looking in the assistant microscope binoculars. As described above, using the microscope is critical for precision retraction, suction, and irrigation. Two surgical suctions are used—one finer than the other—to ensure blood and fluid is cleared from the surgical site and inside the bony crypt. This is the same Luer lock tip, placed on a fine suction tip, used in non-surgical endodontics. Precise retraction of the surgical flap is also performed looking through the binoculars, so that tearing or crushing of the flap is prevented. Wetting the flap, using water from the Stropko irrigator, is done by the first assistant. Some adjunctive tasks that can only be performed whilst working in the microscope may be required, such as soft tissue handling and dissection (Fig. 8.2).

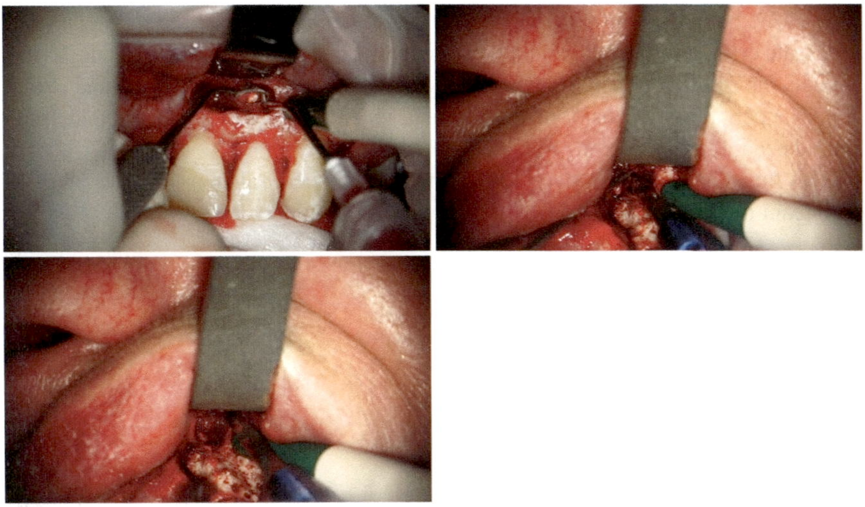

Fig. 8.2 (i) First assistant managing surgical field by placing fine suction tip in crypt and defogging mirror; (ii and iii) performing soft tissue dissection because operator's hands are occupied retracting flap and holding soft tissue lesion

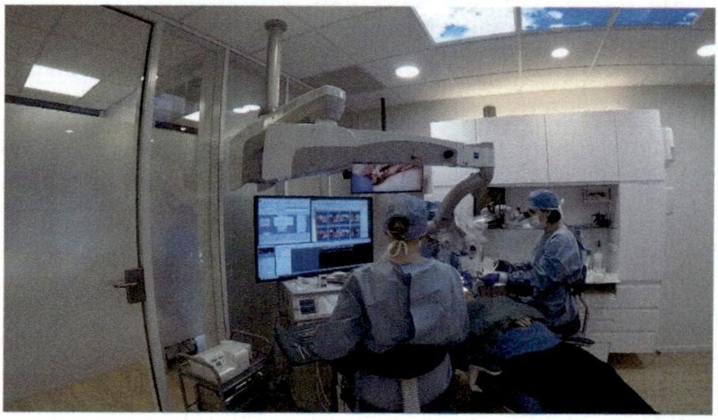

Fig. 8.3 Second surgical assistant following procedure on monitor and managing instruments

Video link to microsurgery overview:

The second surgical assistant, sitting on the right side of a right-handed operator, mainly manages the instruments, materials, and documentation (Fig. 8.3). Because the binoculars are being utilised by the first assistant, a live microscope video feed on a separate monitor allows the procedure to be visualised and the treatment steps to be anticipated. Having this perspective outside the microscope gives a good overview of the patient, the respective stage of the microsurgical procedure and also to anticipate critical steps that cannot be done by the operator and first assistant. These include flap retraction (usually when working on the patient's right side) and irrigation. The dental cart is positioned to face and give easier access to the second assistant, instead of the operator.

This flow chart gives an overview of the main procedural steps and the roles for each team member (Table 8.1):

Table 8.1 Workflow for each microsurgery team member

Phase	First assistant	Operator	Second assistant
Role	Field and crypt management Flap hydration Adjunctive microscopic tasks	Devise and follow surgical plan Manage procedural adaptations Lead the team	Follow and anticipate1-2 steps ahead Prepare instruments and materials ready for the operator Documentation
SETUP	Prepare instruments on starting checklist Chlorhexidine rinse		Prepare instruments on starting checklist Adjust monitor and cart position
LOCAL ANAESTHESIA	LA syringe and suction		
SOFT TISSUE MANAGEMENT	Use suction (larger) and Stropko irrigator for wetting the flap	Flap retraction Bony groove if required	Scalpel and elevator
OSTEOTOMY AND CRYPT MANAGEMENT	Two suctions		Lip retraction depending on tooth Handpiece Curette Haemostatic agent/crypt control material
ROOT END MANAGEMENT	Two suctions Flap hydration Damp gauze	Move patient head to 2nd position Compress flap	Ultrasonic Micro mirror Filling material Radiographs
SUTURING	Two suctions Damp gauze	Move patient head to 1st position Compress flap	Sutures and tying instruments Microscissors
POST-OPERATIVE	Sterile gauze	Provide instructions	Record surgery details in patient record

Before the Appointment

Before the operatory is set up, a microsurgical treatment plan must be devised, which includes any alternative management options mid procedure. If there is a root fracture, what will you do? If there is a perforation or lateral canal, how will it be managed? An integral part of this plan is a high resolution cone beam computed tomography (CBCT) scan. This gives a three-dimensional plan, so that time is not wasted during treatment. Any measurements or analyses of the surrounding anatomy can be done beforehand. The osteotomy site is calculated, ideal amount of root resection measured, and canal angulation determined beforehand, which are important factors related to the treatment outcome.

Pre-operative photos with lips retracted are needed to plan the surgical flap design and soft tissue management, including number and location of sutures. Is a submarginal flap design going to be used? How about a sulcular flap, if there is bony dehiscence at the alveolar crest? Or even, a combination sulcular and submarginal flap? All of these steps should be carefully planned and rehearsed before the procedure—similar to any non-surgical treatment (Fig. 8.4). Without a stored pattern and plan for the team members to follow, it will result in inefficiency, wasted movement and time.

First Surgical Assistant

The instrument setup for the first surgical assistant is minimal compared to for the second assistant. This is because the main role is field management using the assistant-side binoculars, which involves precise suctioning and tissue retraction. Two surgical suctions, one with a fine aspirator tip, are used (Fig. 8.5). The mirror

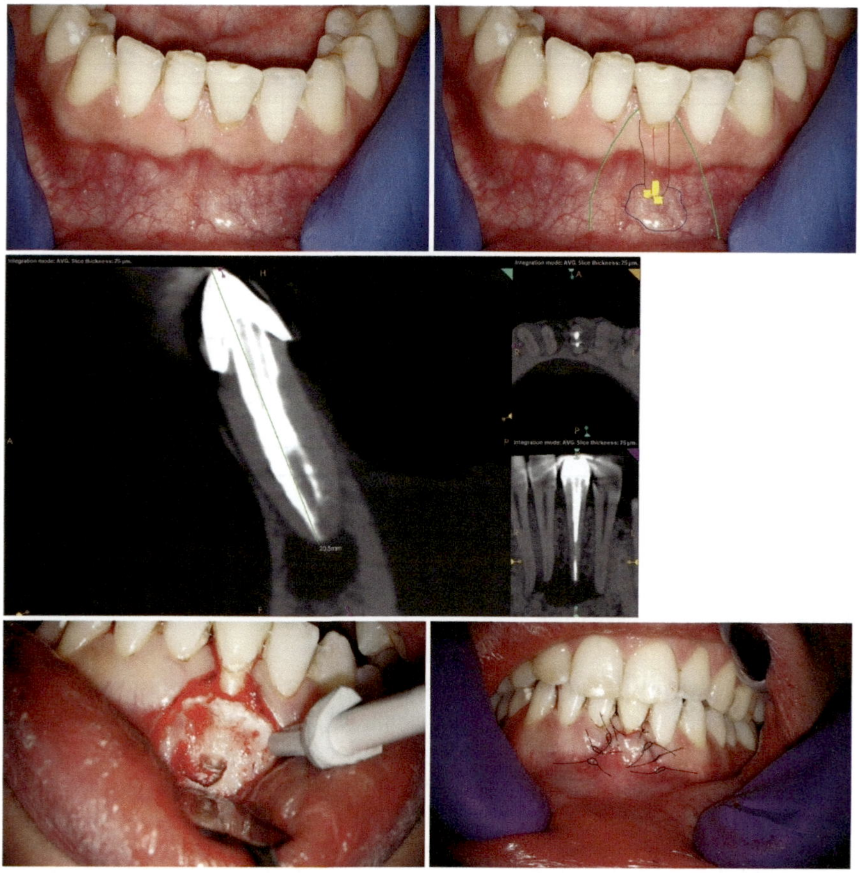

Fig. 8.4 Pre-operative workup for endodontic microsurgery on tooth 31 (i) pre-operative photo, (ii) flap design plan, (iii) CBCT scan showing root length measurement, (iv) completed root resection and root end filling, (v) post-operative sutures, (vi) suture removal appointment, 12-month follow-up, (vii) clinical photo, and (viii) periapical radiograph

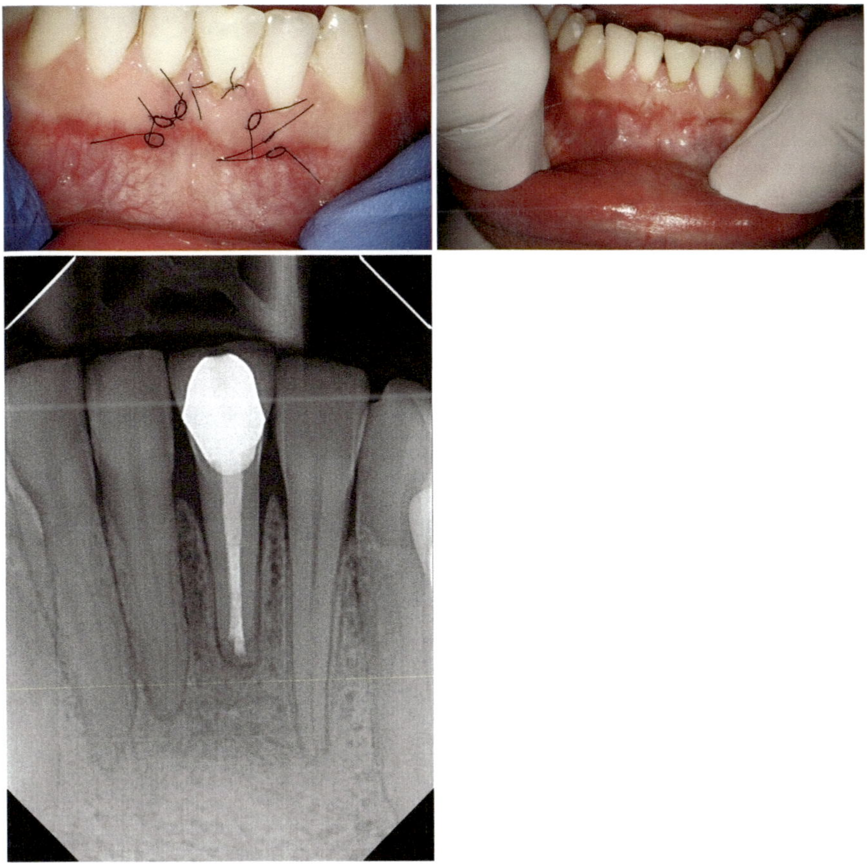

Fig. 8.4 (continued)

and local anaesthetic setup are used, just as for non-surgical treatment. A Henahan retractor (with silicone coating for lip retraction), Carr 45 degree, and Prichard (smaller head) retractor are also on the side of the first surgical assistant. The Henahan retractor is used by the first assistant during suturing. The assistant air/water Stropko irrigator is also set up and ready for use.

Second Surgical Assistant

Most microsurgical instruments and materials are managed by the second surgical assistant. On the dental cart, the setup includes: Stropko irrigator, air-only Stropko, 45 degree high speed handpiece with rear air exhaust and ultrasonic(s) with tip oriented for the tooth in question. The instrument trolley is separate and on wheels to ensure it is in the circle of influence. Setup on this side should follow the principle of keeping it simple and only placing the instruments that will be most commonly used (Fig. 8.6).

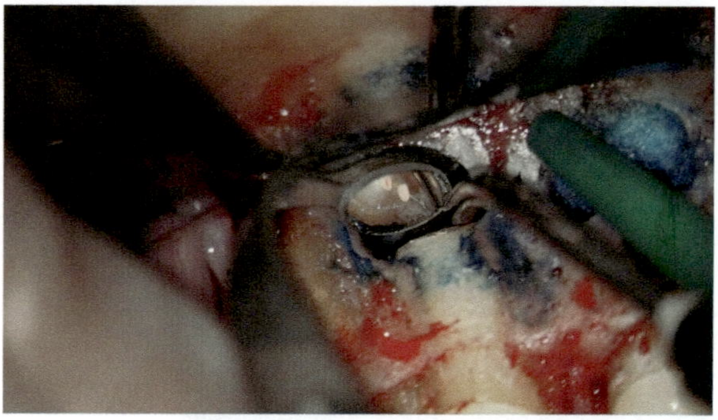

Fig. 8.5 Two suction tips used by first assistant. Fine 18 gauge Luer lock tip placed in crypt and surgical suction tip placed peripherally. This allows visualisation of the isthmus

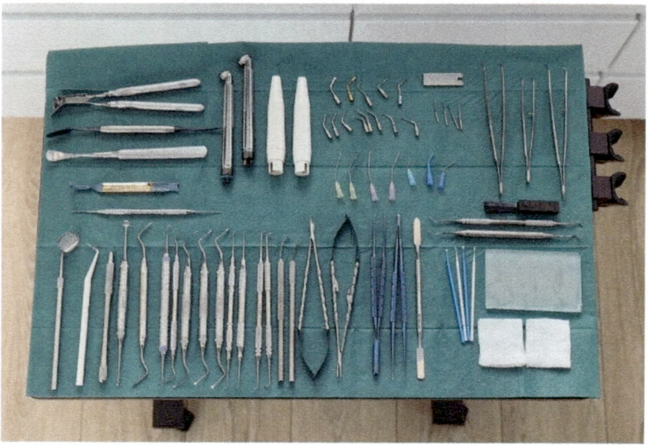

Fig. 8.6 Microsurgical instrument setup

Microsurgical Instruments

One of the most important aspects of successful microsurgery, second to the microscope, is the use of correct instruments and materials. Precision is required in all steps and these instruments need to be designed with this in mind. It is crucial to ensure instruments are used in a stable position, whilst trying to minimise physiologic tremor.

Hand instruments need to be long enough to be held in the pen grip (Fig. 8.7) or external precision grip where the shaft is held *external to the palm* (Konz and Johnson 2008). This is where between the index and middle finger, there is a "cradle" to support the instrument, with the thumb on top for grip and rotation. This tripod-like grip provides stability and if the instrument is too short, the pen grip will lack stability. The middle finger absorbs stress and helps minimise tremor. The apex of the thumb cleft and the side of the index finger provide anti-tremor support, guidance, and feedback from pressure and force. The edge of the ring and pinky finger, and the side of the hand and wrist should have a stable surface to rest on and control tremor (Chan and Velasquez-Plata 2022). The external precision grip is in contrast to the internal precision grip, also called a "motorcycle" grip where the shaft is held *internal to the palm*. This used to be a recommended grip for suturing historically, but is not recommended anymore. In microsurgery, the operator should not be using an internal precision grip for handling any instrument.

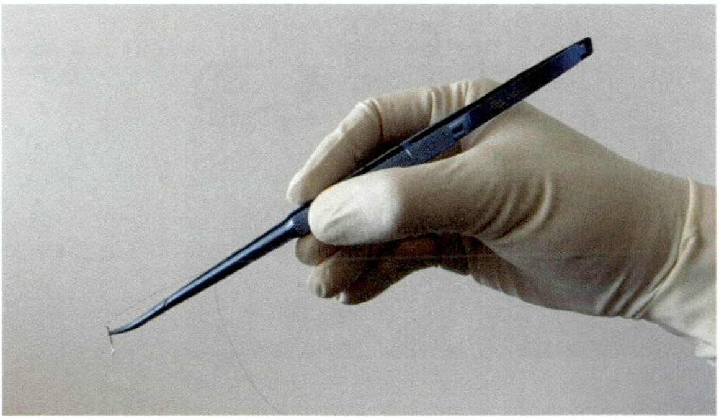

Fig. 8.7 Microsurgical needle holder with Castroviejo lock design, being supported in the "cradle" at the apex of the thumb cleft and the side of the index finger

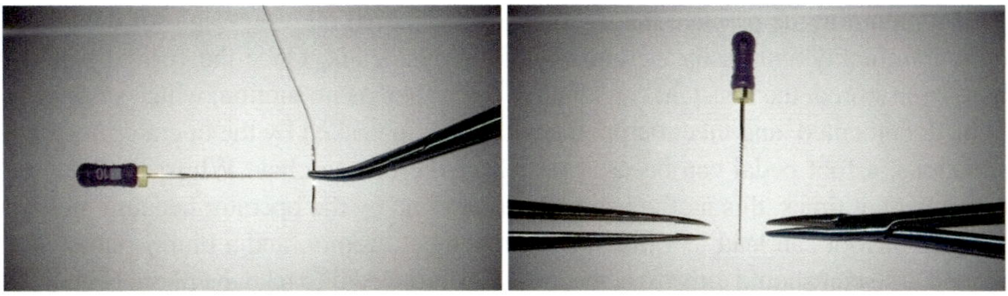

Fig. 8.8 (i) Tip of needle holder in relation to size 10 hand file, (ii) tying tweezers (left) in relation to needle holder (right) and size 10 file in middle

Because very fine suture needles are used, the jaws of the needle holder have to be very small (1 mm maximum width at the tip) so they do not crush the fine needle. The internal surfaces should have a diamond, tungsten carbide or zirconia coating, for roughness, to stabilise the needle securely so it does not slip or rotate in the jaws. A Castroviejo design with a spring lock is recommended because it allows the jaws to be opened or securely closed with a minimal squeeze of the thumb and index finger. A micro needle holder should only be used with small sutures (no larger than 6-0) because a larger diameter needle will warp the delicate beaks and damage the needle holder beyond repair.

Tying micro tweezers should also have a very fine tip (maximum tip width of 0.5 mm) to securely hold the suture material. For both needle holder and tweezers, curved tips are preferable for greater access when holding them in a pen grip (Fig. 8.8). The cross section for hand instruments needs to be of semicircular halves, so that rotation along its long axis can be easily done by using the thumb and index finger. A semicircular design also has a greater margin for slip prevention and greater control compared to a flat surfaced hand instrument (such as a scalpel handle).

Documentation

The second surgical assistant will have a clear view of the operating field through the live feed monitor. All photo documentation is controlled by the *second assistant* operating the mouse. This should be set up so the photo is automatically saved into the

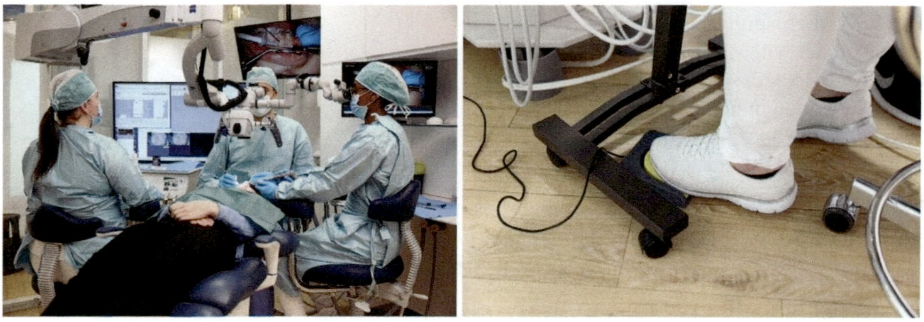

Fig. 8.9 Second assistant following monitor, ready to take photo with (i) mouse or (ii) foot pedal

patient record in the practice management software. All relevant stages of the procedure are quickly and easily captured with a mouse click. Once the correct patterns have been stored, the assistant can identify when there is no motion, which steps need to be documented, and take the photos without being asked by the operator. In some situations, a foot pedal can be used to trigger the mouse click. Whilst this may be convenient at times, this method should not be done by the operator because moving the toes or foot can lead to whole body movement, tremor, and a blurry photo. The second assistant should either use the mouse or foot pedal to take the photo (Fig. 8.9).

List of Instruments

Cart

Stropko irrigator
45 degree handpiece with Lindemann bur
Ultrasonic with tip loaded

First assistant

Large mirror
LA syringe
Periodontal probe
Prichard retractor
Minnesota retractor
Henahan retractor
Carr 45 degree retractor
Suction with 1/4" (6.4mm) tip
Suction with 1/8" (3.2mm) tip and 18ga Luer lock tip
Stropko irrigator

Second assistant

Microblade scalpel (CK-2 or MB69 with semicircular blade)
Molt periosteal elevator
Endodontic explorer DG16
Lucas curette

Jacquette Curette

Root pick

Tweezers – locking

Periodontal hand scaler (x2 depending on type)

Micro mirror

Spatula

Glass Slab

Lee block for MTA

Lee carver

Dovgan MTA carrier

Root end pluggers/condenser (x2 depending on size)

Castroviejo needle holder with curved tips and width <1mm

Tissue tying tweezers x2 with curved tips and width <0.5mm

Microscissors

Suture 7/0 (and 6/0)

Adrenaline soaked pellets

Cotton pellet/Tefla pad

Gauze

Positioning

Operator positioning is the same as for an examination, non-surgical endodontic and restorative treatment, and all other operative procedures—with the operator sitting at the 12 o'clock position. It has been mentioned before that learning a procedure one way is easier and produces better outcomes than learning the same procedure in multiple ways. This is another reason why a swivelling patient chair is required—so that all treatment can be done in the same operator position; only the patient is moved.

Both surgical assistants require a stool with armrests. This provides an optimal muscular rest position and is absolutely critical for the first assistant who would at times be retracting the surgical flap or lip. Stability and precision is crucial in this regard, which cannot be achieved without proper support of the arm holding the retractor.

The first surgical assistant is facing the patient's mouth and looking in the assistant-side microscope binoculars, meaning the position is the same as for all other treatment procedures—sitting in the 3 to 4 o'clock position. Depending on the assistant's arm length, it would be closer to 3 o'clock to ensure the right arm is on the armrest whilst being able to reach the surgical site. This also ensures any instruments on the back wall tray are in the circle of influence.

The second surgical assistant is mainly facing the monitor showing the live feed through the microscope, whilst exchanging instruments and materials. This position is at the 8 o'clock position, using the right arm to exchange instruments with the operator. The dental cart and the instrument tray are facing the second assistant so that all instruments, handpieces, and materials are directly in front and within the circle of influence.

Patient positioning changes depending on the tooth treated and the stage of the surgical procedure. Moving the patient is *the most critical* component in achieving proper ergonomics and efficiency in endodontic microsurgery (Fig. 8.10). This should be explained to the patient beforehand, so they are aware of and can be prepared for the required movements. The patient's head is looking straight for anterior teeth; the head is turned left with the patient chair swivelled to their right for

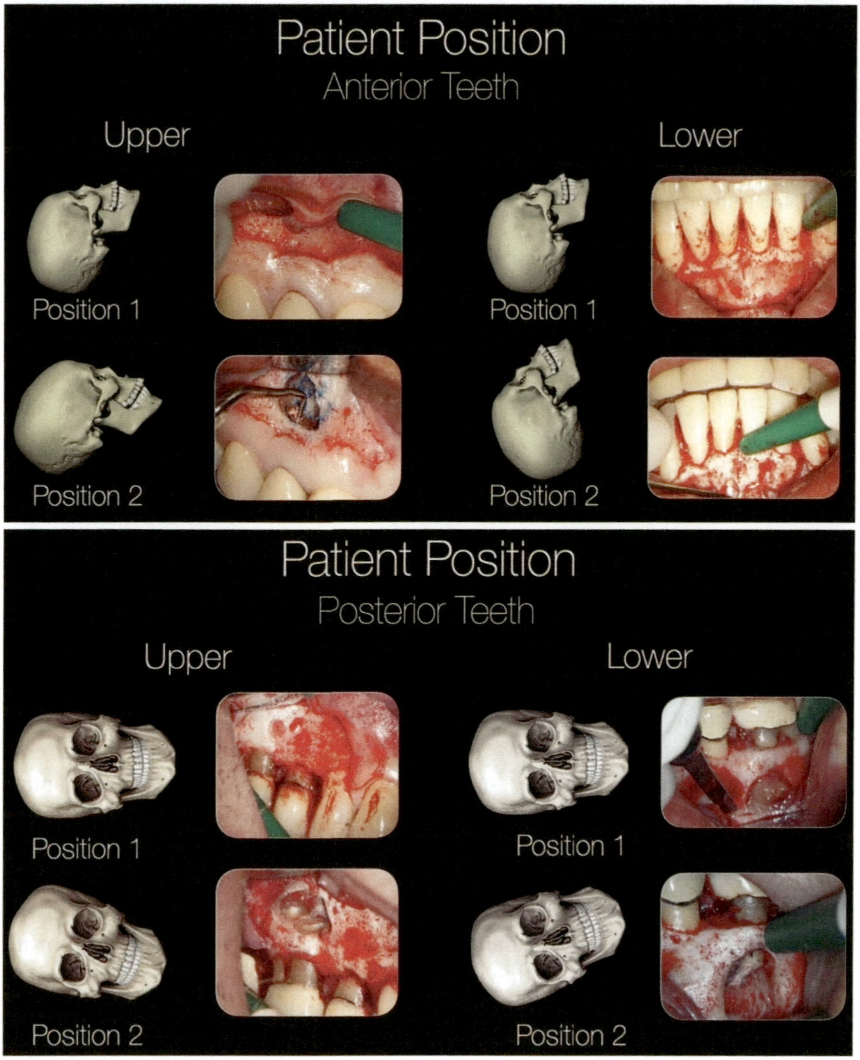

Fig. 8.10 Patient head positions (i) anterior and (ii) posterior teeth, (iii) position 1 for anterior teeth, (iv) position 2 for upper anteriors, (v) position 2 for lower anteriors

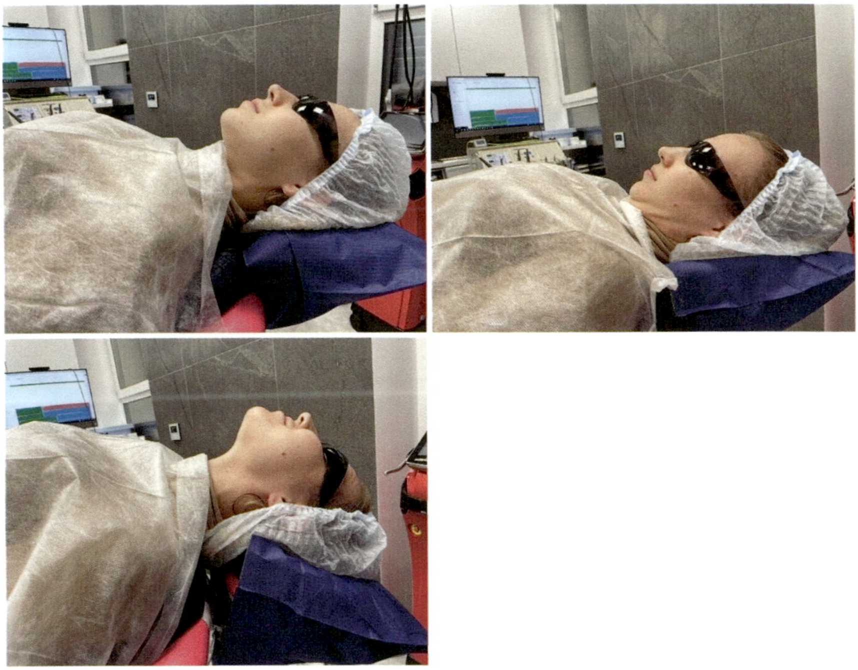

Fig. 8.10 (continued)

right-sided teeth, and vice versa for left-sided teeth. The *first part* of the procedure, starting with LA administration to flap reflection, osteotomy, root resection, and curettage, has the patient oriented with the long axis of the tooth being parallel to the floor. This gives the optimum viewing position and angle for all of these steps, given that the long axis of the microscope is perpendicular to the floor. For a maxillary tooth, this would typically be the same position as for non-surgical treatment. Mandibular teeth would require the patient to tilt their chin up and back slightly, due to the root angulation and incisor proclination.

The *second part* of the microsurgical procedure occurs from inspection and preparation of the canal onwards. From root end ultrasonic preparation to root end filling, direct visualisation of the entire root end is required to make it easier to correctly prepare and fill the canal in the right angle. For this to occur, the patient's chin should be tilted downwards towards the chest for a maxillary tooth. For a mandibular tooth, the opposite occurs with the chin tilting upwards and away from the chest. By showing the patient these movements of first and second position and rehearsing beforehand, they can be ready during the procedure, preventing a delay in treatment. Greater visualisation will also result in better clinical outcomes.

Towards the end of the procedure, when suturing occurs, the head position can be moved back to the first position. The dynamic nature of these patient movements means that the ideal working position can be quickly and easily achieved without moving the microscope and wasting unnecessary time and effort for the team.

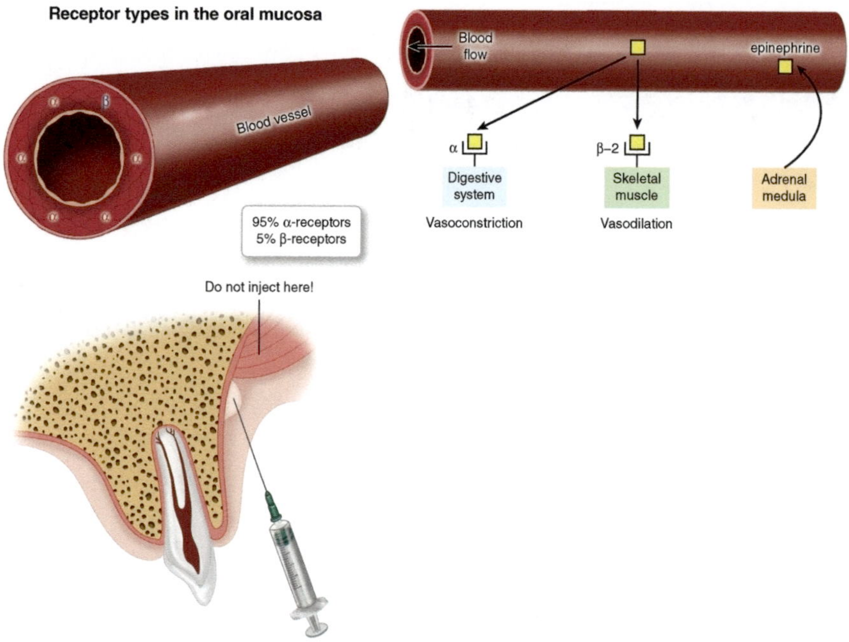

Fig. 8.11 (i) Blood vessel with predominantly α receptors, (ii) α receptor activation leading to vasoconstriction, (iii) schematic diagram showing to inject local anaesthetic in mucous membrane and not into skeletal muscle, which leads to vasodilation

During the Appointment

To minimise wasted patient chair movement, a pre-operative chlorhexidine rinse is given before lying the patient down. Patient protective sunglasses are also given. When lying down, a surgical bib is placed to ensure an aseptic environment over the patient's upper body.

Local anaesthesia is administered with the patient lying down and after the different head and neck movements for the various surgical steps are explained and rehearsed. Just like in the non-surgical Chap. 7, the advantages for microscope anaesthesia are that it is delivered with more precision, less pain to the patient and in a relaxed and neutral position for the operator. You also clearly observe the way the volume of anaesthetic is spreading in the soft tissues to determine areas of distribution. The first surgical assistant exchanges the anaesthetic syringe, just like for non-surgical procedures. It is important to inject the anaesthetic into the mucous membrane, at the border of the free and attached gingiva, where there are predominantly α receptors that cause vasoconstriction. If the solution is injected into the muscle tissue, where there are mostly β-2 receptors, it will lead to vasodilation, poor haemostatic control, and restricted visualisation during the procedure (Fig. 8.11).

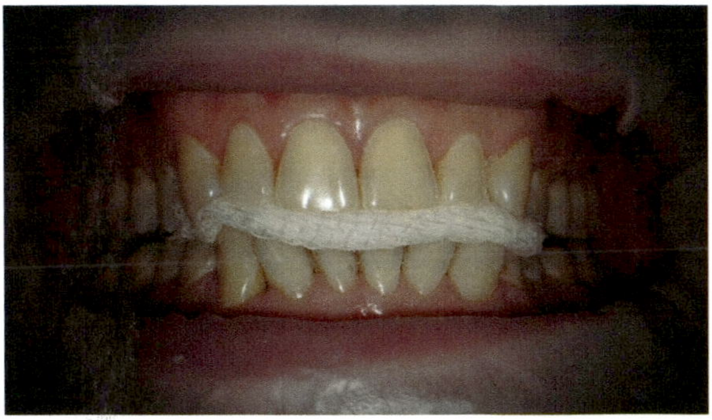

Fig. 8.12 Patient biting on gauze with OptraGate retractor in place, prior to incision

After local anaesthesia administration, to aid in patient comfort, two gauze pieces are passed from the first assistant's left hand to the operator's right hand, for the patient to bite on. A piece is placed on each side of the patient's mouth to help relax the jaw and stabilise the cheek muscles (Fig. 8.12). The Henahan retractor is passed by the first assistant to the operator's left hand. As a general rule just like non-surgical procedures, the first assistant should pass from their right hand to the operator's left hand, or vice versa—left hand to right hand. It is never right hand to right hand because that is more likely to hit the operator's left or mirror/retracting hand in addition to blocking the operator's view through the microscope.

The second assistant passes the scalpel to the operator. Because the position is on the opposite side to the first assistant, instruments are passed differently—from the second assistant's left hand to the operator's right hand. The other difference is that the instrument is held by the assistant at the opposite end of the shank and there is no pulling motion—it is just placed in the pen grip position with the instrument tip inside the field of view (the microscope light). The aim is for the serrated "grip" portion to be placed into the cradle between the operator's index and middle fingers. To exchange an instrument, it is removed with the assistant's right hand and the new instrument placed with the left hand. This is the case for all other hand instruments that are exchanged (Fig. 8.13).

Soft Tissue Management

Incision is performed with the first assistant closely following the scalpel with the suction to immediately aspirate the blood and ensure the visual field is clear. Another task of the first assistant is to periodically wet the flap with a Stropko irrigator to maintain moisture (Fig. 8.14). If the assistant Stropko is used, just like in the non-surgical chapter, it is held in the left hand to ensure the irrigator is close enough to the operating field and the left arm is on the armrest. The right hand holds the suction because it can be held lower down on the handle gaining more reach whilst the right arm is supported by the armrest.

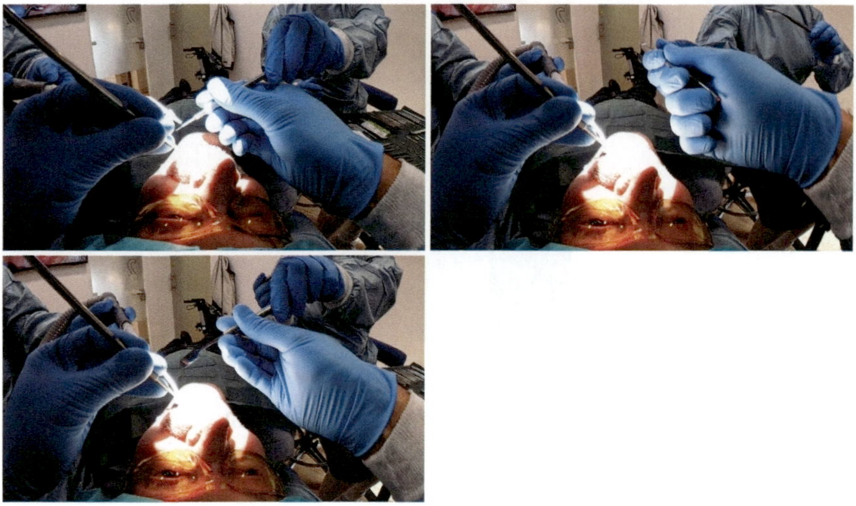

Fig. 8.13 (i) Second assistant passing instrument and (ii–iii) exchanging hand instruments

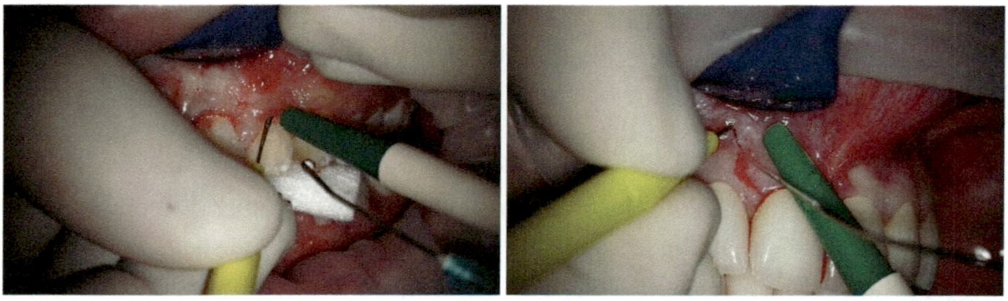

Fig. 8.14 First assistant (i) placing suction close to scalpel during incision, (ii) wetting flap during incision with Stropko irrigator

As part of the pre-operative workup, the flap outline and the cutting direction should be planned. For a right-handed operator, this would often be apical to coronal for the left side vertical incision, left to right for the horizontal or oblique incision, and coronal to apical (or sometimes apical to coronal) for the right side vertical incision. The scalpel should always touch bone the whole time. Incision should be in one controlled, smooth, and continuous motion, as opposed to several start/stop portions along the same incision line. This ensures the flap has a smooth and consistent edge. It will result in less tissue damage and make the flap easier to approximate. For this to occur, a semicircular shaped blade is used, with the three main "edges" used for different cutting movements. Typically used in eye surgery, semicircular ophthalmic blades are designed to cut through soft tissue and can dull quickly when cutting onto bone. Alternative blades that retain edge sharpness for longer are the CK-2 and MB69 semicircular models for endodontic microsurgery. Because only one cutting movement is performed, the scalpel should be in contact with periosteum at all times so that the elevator can cleanly lift the flap (Fig. 8.15).

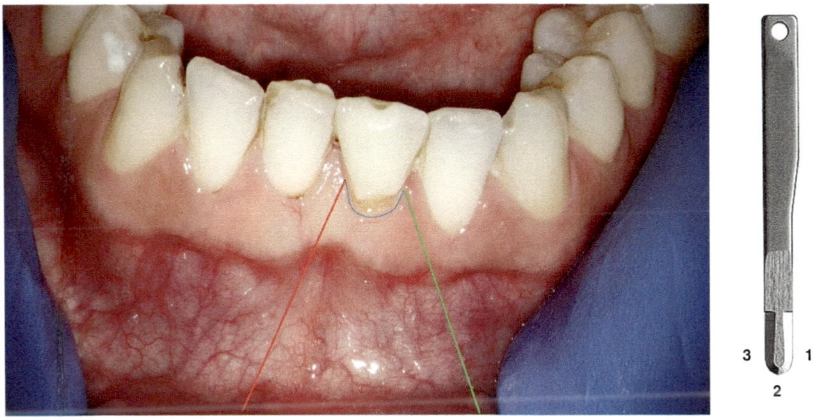

Fig. 8.15 (i) Flap design, with incision order of: (1) green line vertical incision, apical to coronal, (2) blue line sulcular incision, patient's left to right, followed by (3) red line vertical incision, coronal to apical, (ii) MB69 scalpel blade with edge marked number 1 to be used for the first vertical incision, number 2 for the horizontal sulcular incision and number 3 for the second vertical incision

Fig. 8.16 Butt joint vs bevelled incision cross section

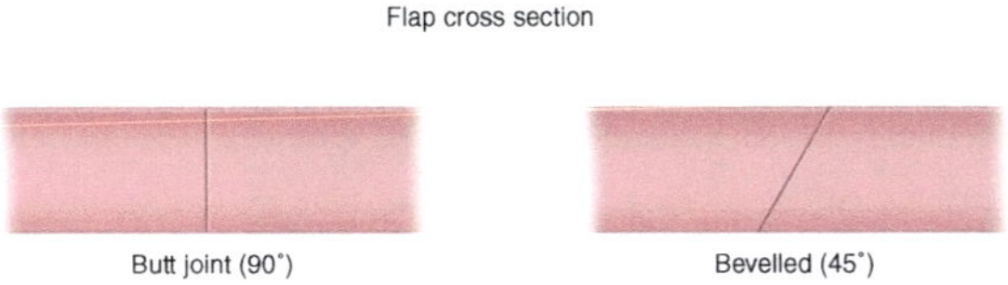

Two types of incisions can be made—the butt joint (90 degree) and the bevelled (45 degree) (Fig. 8.16). We recommend the butt joint incision to help with wound healing, approximation, and suturing. There is a train of thought that suggests the 45 degree bevelled incision can be easier to approximate; however, its design interferes with tissue strength and suture geometry (see suturing section). The same Henahan retractor is still in the operator's left hand. All this is performed with low magnification, so that the complete outline in relation to the surrounding structures can be clearly seen. Using the microscope and the armrests ensures that the operator can work with a stabilised movement and accurately map the incision line, which is for the scalpel to correctly follow the planned outline.

After the incision, the scalpel is exchanged for the Molt periosteal elevator with the second assistant. Magnification is increased so that the operator can see if the entire periosteum is being lifted. This straight instrument is ideal for a majority of situations for flap elevation. It is also more efficient to use one instrument properly than have to spend time changing between using multiple instruments. The wider and sharper end is used to raise the flap, starting from one end of the flap and moving horizontally (Fig. 8.17). The concave surface faces towards bone and the periosteum is lifted from one side to the other. To gain more control, the modified pen grip can be used so the middle finger guides the movement of the elevator. Again, the retractor is still in the operator's left hand. Once the flap is raised and the entire

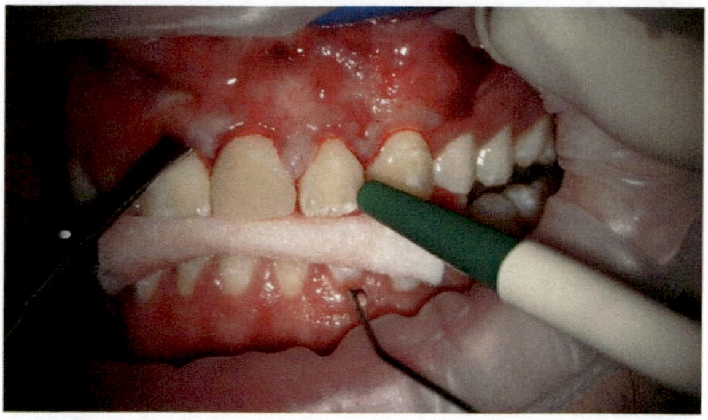

Fig. 8.17 Elevation starting at the patient's right side of the flap and moving horizontally

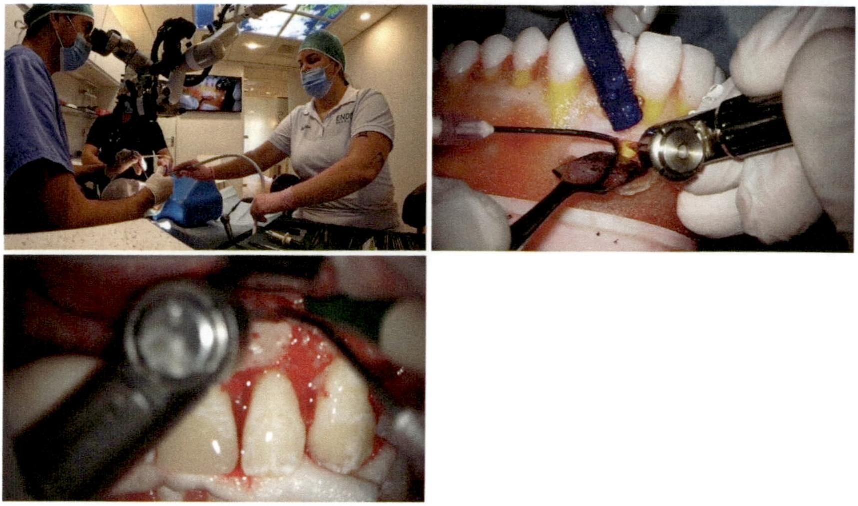

Fig. 8.18 (i) Handpiece being passed to operator (ii and iii) Luer lock suction tip placed next to bur, with larger suction slightly further away

surgical site can be visualised, the next step of performing the osteotomy can proceed. The elevator is returned to the second assistant.

Osteotomy

The second assistant passes the handpiece to the operator (Fig. 8.18). The first surgical assistant uses two suctions during osteotomy. The smaller suction with the fine Luer lock tip is placed right next to the bur. The larger suction is placed slightly further away. The armrests ensure that each hand holding a suction tip is supported, to ensure stability of placement and comfort for the assistant (Fig. 8.18). If there is a bony fenestration, the planned osteotomy site will only require extension of the existing bone loss. If there is a thin layer of bone but no fenestration, the explorer or scaler should be used to manually pierce through the bone in order to check the estimated root tip position. The magnification should be changed to the lowest setting, to have the widest field of view and a better understanding of where the osteotomy will be cut, in relation to the root angulation and estimated root tip position.

Retraction of the flap is done with the Carr 45 degree retractor in the operator's left hand. This should be positioned slightly apically to the planned osteotomy site. Because the armrest is stabilising the forearm, the retractor can be placed on bone without any major risk of it slipping or crushing the flap. This retractor is also serrated to help with instrument purchase on bone. The use of a groove, cut into bone for the retractor to sit in, is an optional step that can be done for those gaining experience—however, it will add more time and wasted movement; the retractor is already stable without the groove because the arm is supported by the armrest.

Root End Management

The second assistant passes the 45 degree handpiece to the operator's right hand, whilst ensuring the cord does not get tangled. It should have the Lindemann bur loaded, chosen because of its bone cutting efficiency. The main reasons for the 45 degree handpiece are that the air is evacuated to the rear connector, removing the risk of air emphysema, plus the angle allows the operator to hold it in a more comfortable position. Of note is that the operator should be careful not to hold the 45 degree handpiece in an angle similar to a conventional handpiece; otherwise, it will cut with a bevel to the long axis of the root. The bevel can lead to inadequate access to a lingual canal and insufficient root resection. This is the value of low magnification, which allows you to clearly visualise the relationship of the bur to the entire root. The same bur is used for cutting bone and root resection, preventing the need for a bur change and wasting a movement and valuable time. An aseptic water supply from the handpiece ensures an aseptic cutting field and removes the need for the second assistant to irrigate with saline. Although every microsurgical scenario will be different, one principle to follow is to remove enough bone to visualise the root tip first, then resect the root. Initially, the magnification will be low during osteotomy. When root resection is started, the magnification should be increased to a medium level to ensure the correct length of root is being removed. After some initial cutting 2–3 mm into the root surface, the magnification should be changed back to the lowest setting to ensure the resection occurs at 90 degrees to the root long axis with no bevel angle in any dimension. This is recommended as opposed to shaving the root tip gradually—it wastes time and removes the reference of the estimated length of root to be removed (Fig. 8.19).

After root resection, curettage occurs with the second assistant exchanging the handpiece with the Lucas curette. The first assistant uses the same two suctions. Granulation tissue can be lifted out of the crypt by facing the concave side towards the internal crypt surface and scraping along the bone towards the posterior. This frees the granuloma from any attachment to the crypt wall. Both left and right sides of the Lucas curette can be used with the operator rotating the instrument 180 degrees

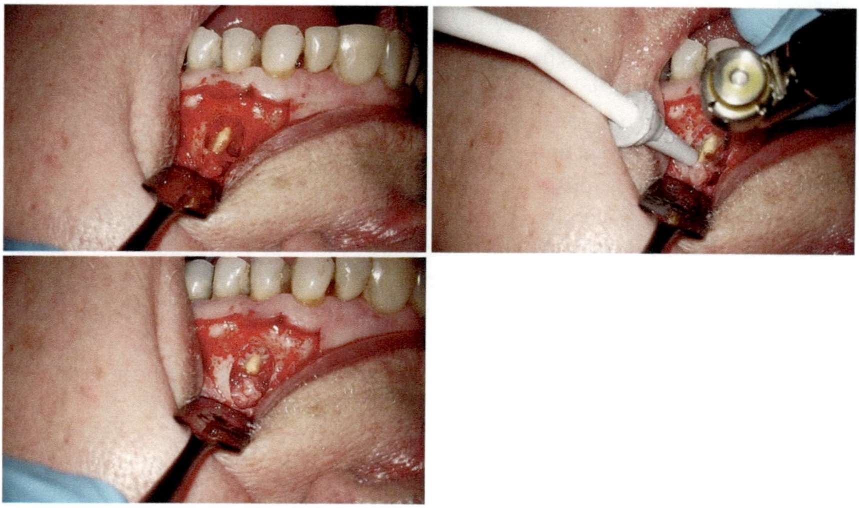

Fig. 8.19 (i) Exposed root tip prior to root resection, (ii) partway through resection, (iii) root resection completed

with only finger movements. An anti-clockwise motion should be used, so as not to confuse the assistant with the clockwise motion that starts the 1–2 movement, for a right-handed operator. In some situations, the root tip will need to be removed with a root pick, especially if access in the crypt is limited. The granulation tissue is removed by the second assistant exchanging the curette with tweezers, which then is placed in a specimen jar with formaldehyde for histopathological analysis.

It is possible that curettage may be required before root resection to aid in visibility and to prevent soft tissue being caught in the bur. If so, the exchange between curette and handpiece is done efficiently, to the point the second assistant is likely to anticipate this requirement by watching the monitor showing the microscope live feed.

Once the required amount of root has been resected and soft tissue curettage completed, inspection of the root end occurs. Methylene blue dye is used to stain organic tissue and tell the operator which areas need further management, such as the periodontal ligament, a canal isthmus or organic tissue in a lateral canal foramen. It also can easily identify a vertical root fracture. A drop of the dye is placed on a microbrush or cotton pellet and passed by the second assistant to the operator's right hand. To remove the excess, the Stropko syringe is then exchanged to lightly wash and dry the surgical site. Alternatively, the first assistant can use the assistant Stropko or Luer lock suction tip, getting it ready whilst the methylene blue is being used. Invariably with good magnification and field management from the first assistant working through the microscope, you can see the fracture clearly without the need for methylene blue dye (Fig. 8.20).

To aid in haemostasis and to prevent gutta percha or other material from entering the crypt, an adrenaline soaked pellet in locking tweezers is placed at the back of the

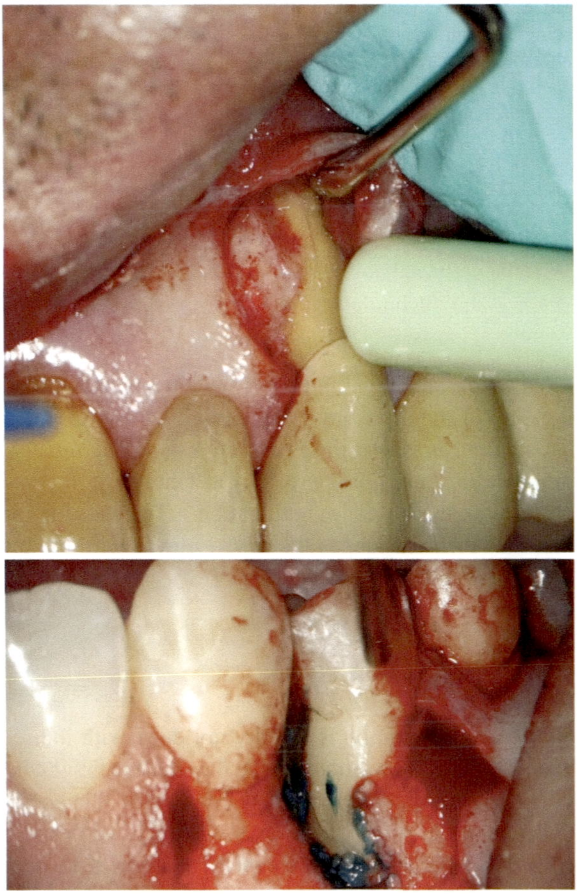

Fig. 8.20 (i) Vertical root fracture, (ii) with methylene blue dye

crypt. Alternatives include a collagen sponge (CollaCote, Integra Lifesciences Corporation), which aids in clotting because collagen is an activator of the intrinsic pathway of the coagulation cascade. Oxidised regenerated cellulose (Surgicel, Ethicon Inc) can also be used because it provides a matrix for platelet adhesion and aggregation. Telfa pads (Kendall) are non-adherent cotton pads that do not stick to the wound. These are all cut to size and placed in locking tweezers. All of these exchanges occur with the second assistant, whilst the first assistant is still managing the field of blood and moisture with both suctions. This all occurs with the lowest magnification.

Inspection is the stage where patient positioning changes, so that root end management can occur with direct vision. As has been previously rehearsed with the patient prior to the procedure, for a maxillary tooth, the patient's chin is now tilted downwards towards the chest. For a mandibular tooth, the chin is instead tilted upwards and away from the chest. The magnification is now changed to a high level, so that inspection with the micro mirror—exchanged with the second assistant—can occur and all the relevant details are seen. The root outline, the location of the

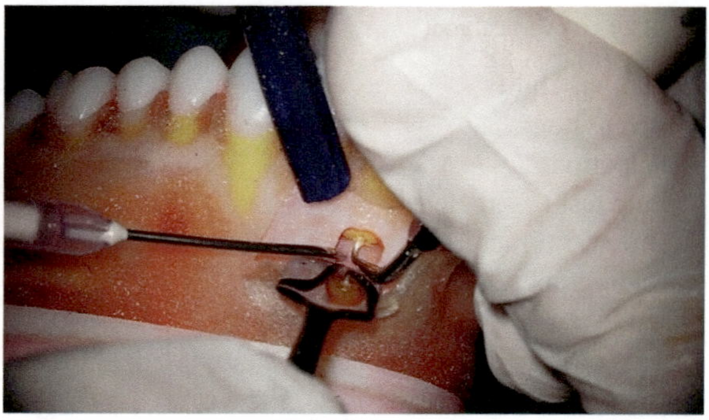

Fig. 8.21 Two suctions being used during ultrasonic preparation

filled canals and/or isthmus, and any lateral canals are visualised in order to plan the location of the ultrasonic preparation. During inspection of the root end, the first assistant ensures that the fine suction is inside the crypt to defog and prevent any moisture dirtying the micro mirror. Once high magnification inspection has been performed, the lowest magnification is selected for canal preparation. The second assistant then exchanges the micro mirror with the ultrasonic handpiece with the correct tip already loaded.

Canal Preparation

Ultrasonic preparation is usually not done with indirect vision; the operator's left hand should still be retracting the flap and the first assistant should still use two suctions. With the patient in the second position, the operator should be mindful of the updated angle of the long axis of the canal when removing the gutta percha (Fig. 8.21). For a maxillary tooth, there might be the tendency to point the ultrasonic tip towards the coronal portion in an upwards (buccal) direction. This will not only lead to an iatrogenic error by preparing the buccal surface of the canal and not the canal itself, but possibly lead to fracture of the ultrasonic tip. Again, this is the value of low magnification, where you can see the long axis of the entire tooth. Because the ultrasonic is cutting through soft gutta percha or empty canal space, only light pressure is required. If there is a lot of cutting resistance, then the angle of the ultrasonic tip should be checked to ensure it is not cutting off axis into dentine.

Another reason for iatrogenic error in ultrasonic preparation is inadequate haemostasis. If the bleeding is obscuring the surgical site, the bleeding source should be determined and accounted for, whether it is another adrenaline pellet, or more precise suction placement. Once the canal has been prepared, the microscope is changed to high magnification and the ultrasonic exchanged with the micro mirror to inspect the root end cavity preparation. This is to check that the right dimension has been prepared, that the existing gutta percha has been removed and no material is left on the walls. The air-only Stropko irrigator is also used to dry the prepared canal space to check for any filling material remnants on the buccal or palatal walls. By reducing the air pressure, this helps prevent any blood spraying into the canal

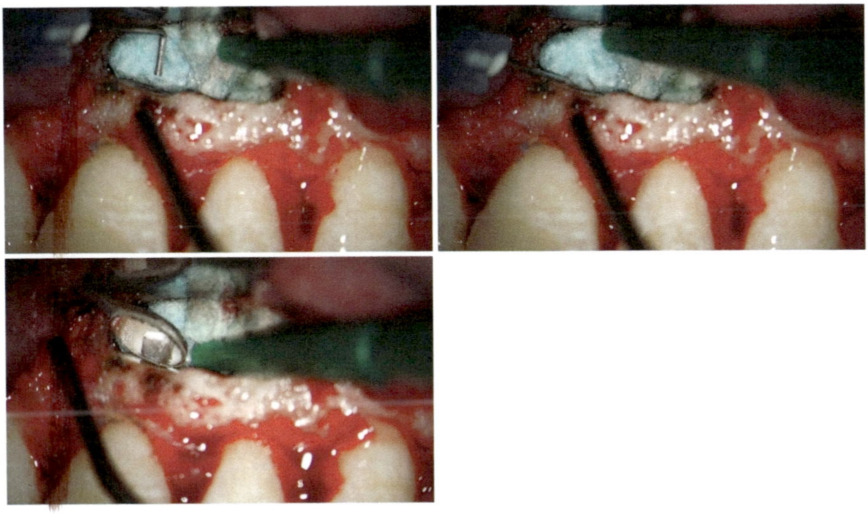

Fig. 8.22 (i) Drying the canal with an air-only Stropko irrigator, (ii) water droplet being evacuated then suctioned with fine surgical tip in crypt, (iii) inspection of dried canal walls

and obstructing vision (Fig. 8.22). Limiting blood into the canal also occurs with the Luer lock tip on the surgical suction. The smaller diameter reduces the amount of blood evacuated, reduces aerosol and provides precise blood removal.

Canal Filling

Root end filling is done by changing the microscope back to low magnification and the material delivered straight into the canal. This is all exchanged with the second assistant, with the first assistant still keeping the crypt and surgical site dry with both suctions.

There are two ways to deliver the root end filling material, which is likely going to be mineral trioxide aggregate (MTA) or a similar putty-type material: using a carrier, or placed using a hand instrument. Either method is prepared by the second assistant, with the first assistant still keeping the field dry with both suctions. Using the carrier method, the material is loaded in the Micro Apical Placement (MAP) carrier and the sides are wiped clean. The second assistant passes it similar to a local anaesthetic syringe to the operator's right hand. This is then exchanged with a micro condenser and used like a hand instrument. This is repeated until the canal is filled to the resected root surface. For many teams, this is a more predictable technique (Fig. 8.23).

The hand instrument method uses an MTA block, where the material is placed into grooves and carried with a Lee carver. The second assistant passes the Lee carver just like any hand instrument and holds the MTA block, with material loaded, into the operating field. This allows the MTA to be taken off the block with the

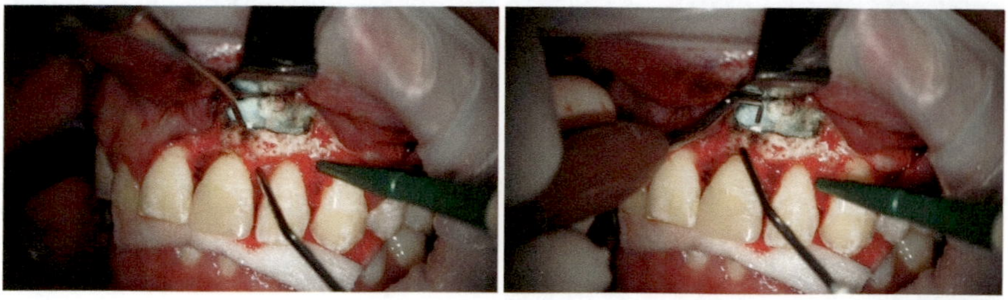

Fig. 8.23 MAP system material (i) placement with carrier and (ii) compaction with condenser

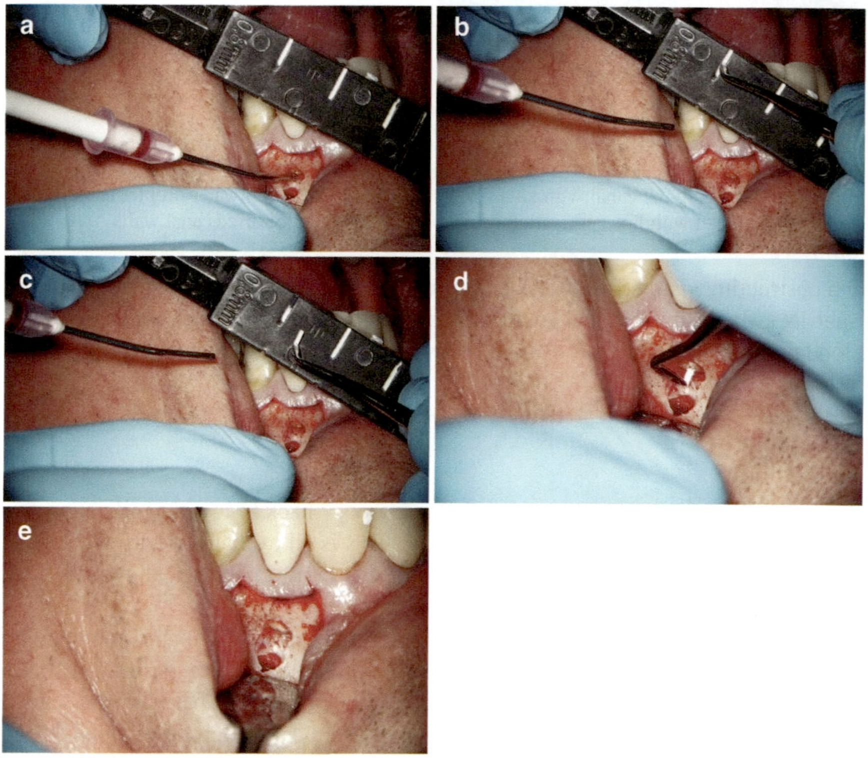

Fig. 8.24 Lee block (**a**) loaded with bioceramic putty material, (**b**) Lee carver right before loading, (**c**) material on Lee carver, (**d**) material right before canal placement, (**e**) filled root end with direct vision

carver and placed into the canal—all with the operator looking in the microscope. Similar to the carrier method, this is then exchanged with the micro condenser with the second assistant and repeated, until the canal is filled with MTA (Fig. 8.24). More precision is required for the second assistant using this technique. A moist cotton pellet in locking tweezers is then used to smooth and clean the filled root surface.

Prior to suturing, the filled root end should be inspected under high magnification to ensure the material is well adapted and the interface with the dentine surface is smooth (Fig. 8.25). If the operator has been careful with material placement and handling, then minimal or no excess material should be in the crypt. This can be removed at the time by the first assistant using the micro suction, or after material placement when the operator uses locking tweezers to remove the adrenaline pellet. Right before the sutures are placed, a check PA radiograph is taken to check for material placement. Once this has confirmed the filling location and density, tissue compression with damp gauze for 15 min should now be performed. Flap compression is done twice—before and after suturing. The first time is to ensure proper wound margin approximation, flap hydration, and accurate suture placement. The second time ensures flap stabilisation, haemostasis, and minimising clot thickness to aid in healing.

Suturing and Wound Healing

Close approximation of tissues and secure closure of wound margins is critical for tissue stability whilst healing. Sutures should maintain apposition of wound edges until scar tissue has sufficient strength (Macsai 2007). In addition to facilitating blood clot maturation and protection of the surgical site, hermetic wound closure by primary intention helps prevent scar formation. Understanding the mechanism of how sutures keep tissues approximated leads to a greater appreciation of how placement should occur. Most importantly, suturing of a passive flap should occur with minimal tension on the wound margins. The tissue surface, tissue thickness, wound margin, and wound surface describe components of the wound that the suture passes through (Fig. 8.26).

Suture Types

A properly placed suture withstands functional forces that open the wound, without impeding healing. Because the suture needle pierces and passes through tissue during placement, the type of needle being used and its direction of travel in tissue has

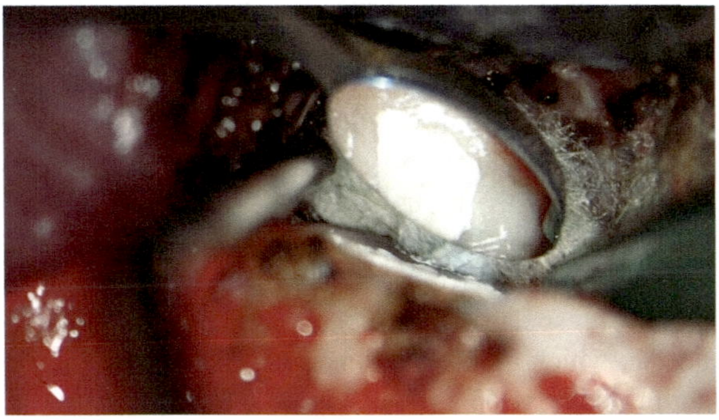

Fig. 8.25 High magnification inspection of root end filling

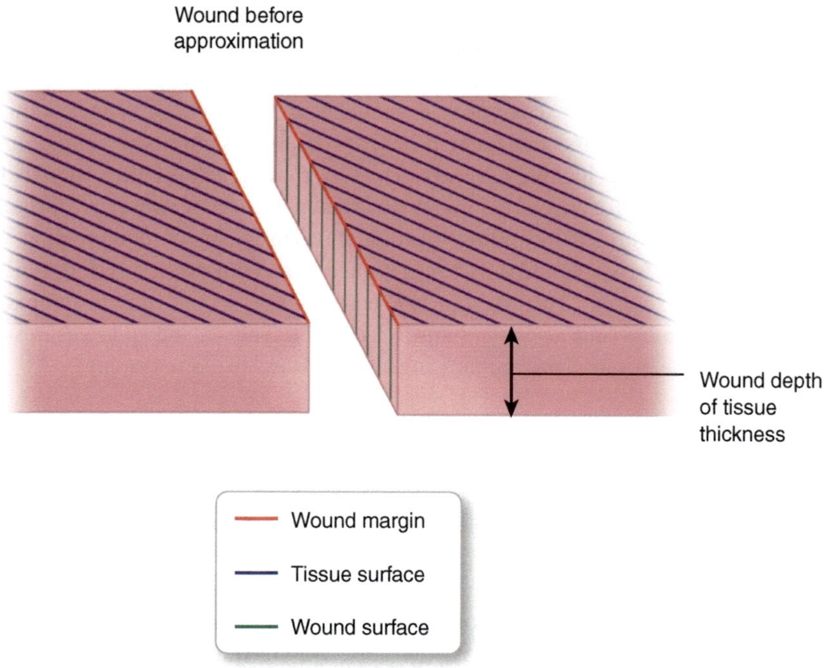

Fig. 8.26 Wound components

to be carefully considered. Microsurgery makes minimal tissue trauma, precision suture placement, and passive wound closure all possible. Small diameter suture needles allow for this to occur. Every needle has three components, which determine which part of the needle should be held by the needle holder. The point should be sharp enough to cleanly pierce the tissue with minimal resistance. If the needle holder grasps the point, it may get twisted and be unable to pierce tissue precisely. The swage is where the suture stand is attached to the needle. Because this is hollow, the needle holder should not be held here because it will deform and possibly detach the suture to the needle. Another reason for not holding at the swage is the needle can be twisted and be unable to precisely pierce tissue. The body is the area between the point and swage. This is where the needle should be held. Needle arc radius determines the needle body length. Most common needle arc sizes in endodontic surgery are 3/8 or ½ circle, corresponding to 135° and 180°, respectively. They provide enough length to pass through tissues, without being too long as to impede movements in areas with limited access. The needle cross section should be determined by the tissue biotype. A cylindrical or taper point cross section is less likely to tear tissue in thin biotypes than traditional or reverse cutting needles, which have a triangular cross section with sharp edges on the concave or convex needle surfaces, respectively. A thicker tissue biotype would benefit from a reverse cutting needle, which reduces tearing when the needle pierces tissue. This is because the cutting edge is on the outside of the needle curve direction, meaning the inner

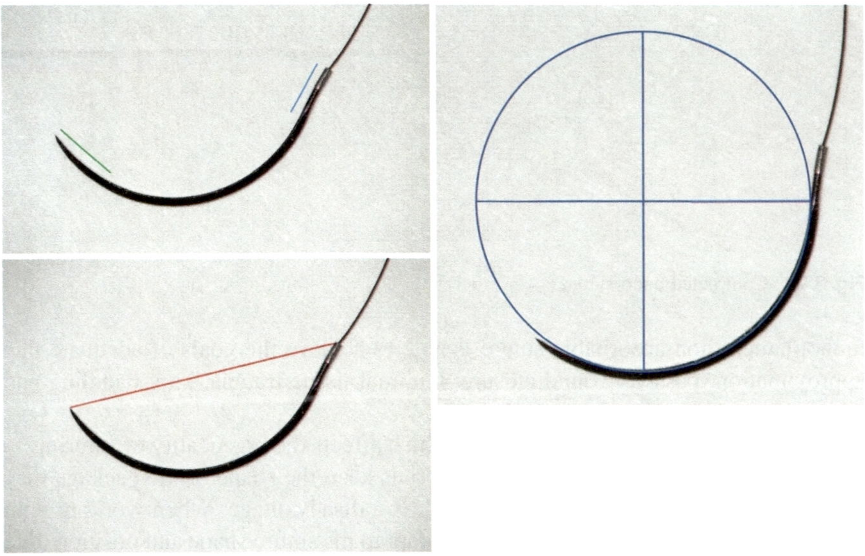

Fig. 8.27 (i) 6/0 suture needle showing point (green line area), swage (blue line area). The body is the section between these two areas. (ii) curvature reference showing a 3/8 circle (iii) red line showing chord length

circular edge of the needle does not apply a cutting force as the needle is passed in a circular motion. In most settings requiring endodontic microsurgery, a thick biotype tissue is not commonly encountered. Chord length is the distance from the needle tip to the swage, which determines the distance the needle can reach from the entry and exit points on the tissue surface, or once passed through tissue surface to wound margin and vice versa (Fig. 8.27).

Suture materials should retain strength throughout the wound healing process, elicit a minimal tissue reaction, and be handled and tied in a manner that can adequately secure the knot. Classification of suture materials can include: monofilament vs multifilament, or absorbable vs non-absorbable. The aim of microsuturing should be emphasised again, which is to impart minimal tissue trauma and passively approximate tissue.

Suture cross sectional thickness is classified in zeros, with more zeros indicating a thinner suture material. A 5-0 (00000) suture is finer than a 3-0 (000) suture. To ensure minimal tissue trauma, passive wound closure and adequate suture stability, micro sutures of no larger than 7-0 should be used. With up to a 5N force, the 7-0 sutures lead to thread breakage rather than tissue breakage (Burkhardt et al. 2008). This is important because a larger suture at the same force is more likely to tear through tissue, resulting in irreversible damage and worse healing outcomes.

Monofilament sutures consist of a single strand, which produce less friction and trauma whilst passing through tissue. They are also less prone to microbial colonisation and inflammation. Multifilament suture materials can be braided or woven, which cause more tissue trauma and a greater inflammatory response. A

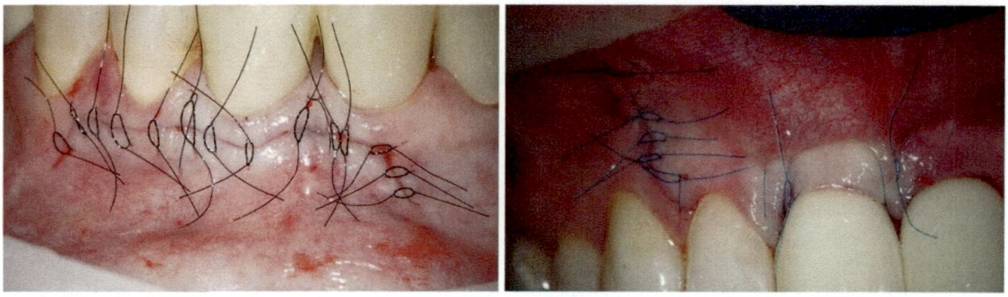

Fig. 8.28 Completed microsutures

monofilament non-absorbable suture is able to achieve the goals of adequate flap approximation, passive wound closure, minimal tissue trauma, knot handling and tying, and strength retention.

One factor not commonly discussed, which affects the practicality of suturing, is the material "shape memory". Some materials keep the shape of the package they are wound in—often an oval shape, which is a disadvantage. When working with higher magnification, the operator has to adapt to the suture shape and position (due to the memory) which restricts freedom of movement. Some materials do not "bounce back" to their original position—typically a treated polyamide material—and stay in the position they are placed, which helps keep the suture in the field of view and is advantageous in tying microsurgical knots.

The ideal suture for most endodontic microsurgery situations is a 7-0 monofilament, non-absorbable suture, with a taper point needle cross section, 9–11 mm chord length and 3/8 circle arc, in either a polypropylene, polyamide, or nylon material.

Suture Geometry

By producing forces with many differing vectors, the suture creates zones of compression and tension. An interrupted suture is the most commonly used suture in endodontic microsurgery (Fig. 8.28). Besides being able to be used for all flap designs, they are independent of each other so if one becomes untied or breaks, the other sutures are unaffected. Most importantly, placement should occur with class 1 or 2 movements.

Unlike suturing in other parts of the body involving only soft tissue layers (i.e. ophthalmology and dermatology), suturing in endodontic microsurgery is mucoperiosteal—meaning soft tissue is approximated on top of hard bone. This

Fig. 8.29 Suture plane showing compression vectors

influences the suture geometry and method of placement because the needle pass direction has to take into account the position of the bone. A single interrupted suture has a suture plane, which is the flat surface of a closed suture loop when adequately tightened (Fig. 8.29). Because this closed loop traverses the wound surface and wound margin, both sides of the wound exert a force onto the suture, which imparts an equal and opposite force to the tissues—i.e. Newton's third law. By understanding this, it can be used to properly plan suture placement.

Suture geometry should be designed so that there are equal forces from the suture to all aspects of the tissue and wound surfaces. This can only occur with a butt joint incision. The interrupted suture has multiple force vectors, which determine how it should be placed to ensure forces are balanced and lead to stable tissue approximation (Macsai 2007). The force directions are: perpendicular to the tissue surface, perpendicular to the wound surface, parallel to the wound margin, and parallel to the suture plane (Fig. 8.30).

Tissue thickness, or wound depth, is a fixed distance which determines the distance from the wound margin to the base of the wound. The suture entry point on the wound surface should be at the base of the wound—to ensure the tissue has enough strength to resist the suture pulling through the tissue, which causes unnecessary tissue tearing. Therefore, the distance from the wound margin to the suture entry point should be equal on both the wound surface and the tissue surface. This ensures an equal force at the wound margin, which prevents wound eversion. This distance should therefore be the same for both sides of the wound, which is termed the "bite size" (Fig. 8.31).

An equal bite size on both sides of the wound means the force is even, and there is no displacement of one side of the wound due to an asymmetrical suture force. For the width and height to be the same, the angle of entry should be 90 degrees to the tissue and wound surfaces. If the angle of entry is not at 90 degrees, this will lead to some adverse events. If it is greater than 90 degrees at the tissue surface, measured from the wound margin, the suture will be too shallow relative to the width. This will lead to tissue tearing or uneven displacement on that side of the wound. If it is less than 90 degrees at the tissue surface, the base of the suture will have a wider

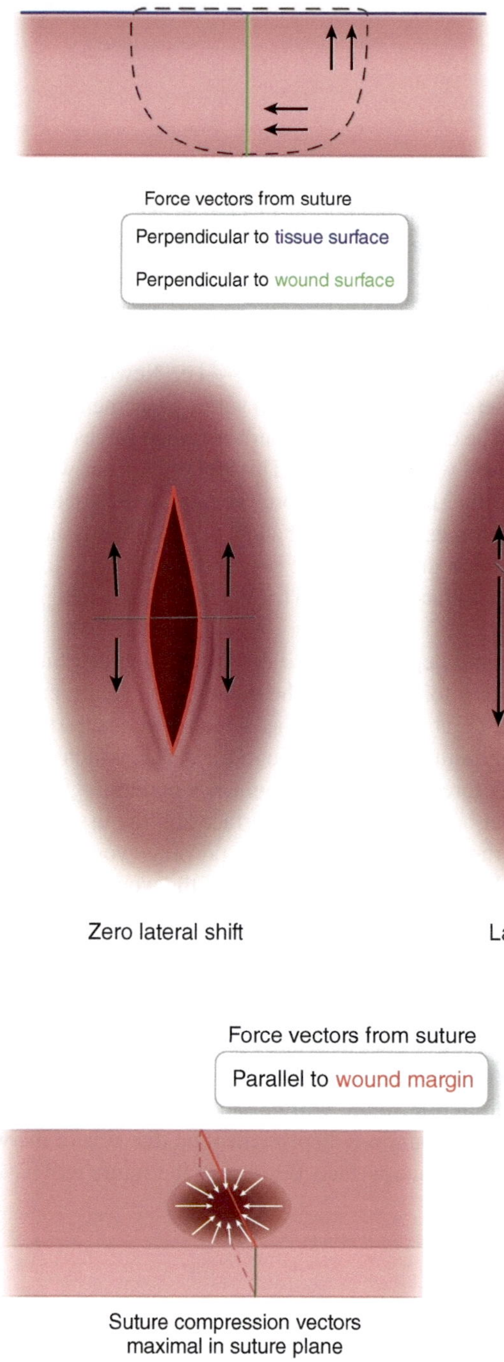

Fig. 8.30 Force vectors (i) perpendicular to wound and tissue surfaces, (ii) parallel to wound margin, and (iii) parallel to the suture plane

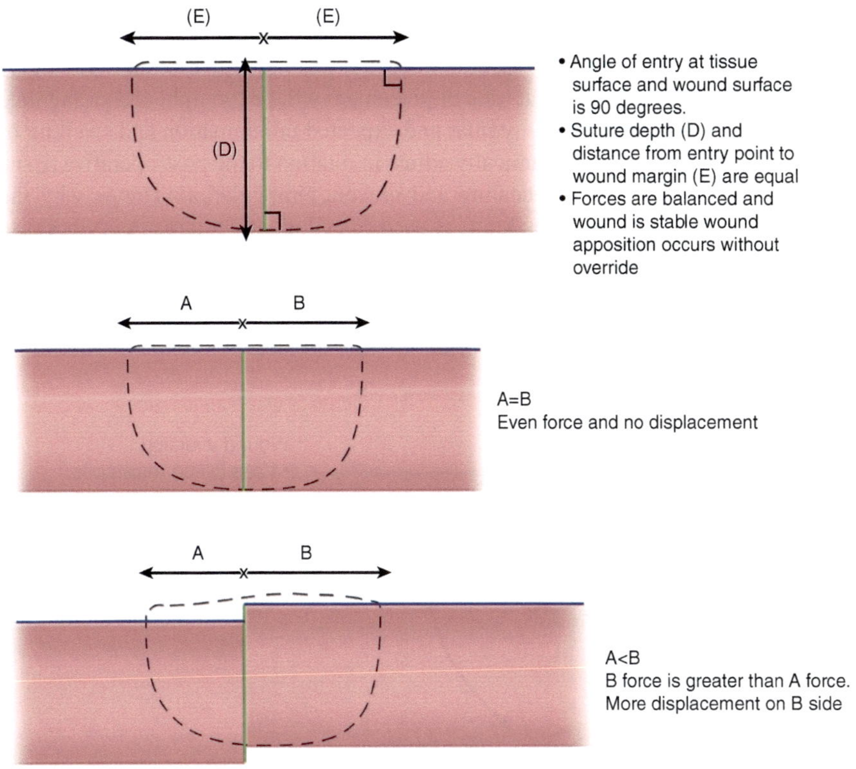

Fig. 8.31 The ideal angle of entry at the tissue and wound surfaces is 90 degrees. The bite size (E) on both sides of the wound should be equal to ensure stable wound apposition. Wound displacement occurs if bite sizes are uneven

distance from the wound edge than at the tissue surface. This will lead to eversion of the wound (Fig. 8.32).

A force vector parallel to the wound margin occurs when the suture is placed obliquely to the wound margin. This results in a lateral shift of the approximated wound edges. To prevent this sideways movement, the suture should be at 90 degrees to the wound margin or incision line. This ensures the vectors parallel to the wound margin are equal on both sides and cancel each other out. With this in mind, the needle pass, or direction of entry and exit at the tissue surface, should always be 90 degrees to the wound margin (Fig. 8.33). Primary closure can therefore occur with the suture perpendicular to the incision line. This is not only for the initial needle pass, but also for tying the suture and closing the knot.

Suture tension can be defined as the difference between the tied suture circumference and the ideal suture length. If the tied length is tied too tight, the suture tension is too high and tissue strangulation will occur. The opposite occurs when the suture circumference is too large and tissues are too loosely approximated. This will lead to healing by secondary intention at the wound margin. It should be emphasised that suture tension should take into account post-operative tissue swelling. A tightened knot of ideal tension at time of placement will be too tight and strangulate tissue after 1–2 days because of regular and expected inflammation and swelling of the tissues. If the knot can dynamically adjust in relation to the post-operative tissue changes, that will assist with healing and prevent strangulation. This is why the square knot with space between knots is used in endodontic microsurgery because the first loop loosens in response to tissue swelling, which tightens the knot in relation to the second loop.

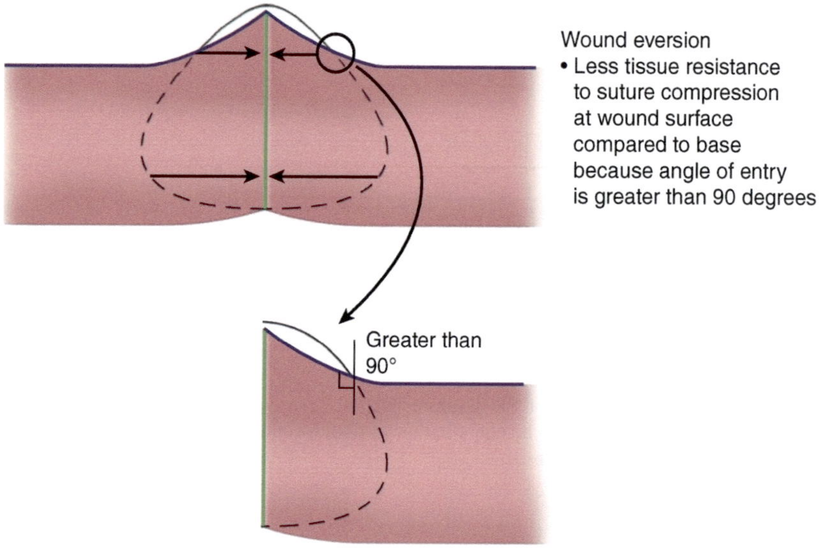

Fig. 8.32 Wound eversion occurs when the angle of entry is greater than 90 degrees at the tissue surface

Fig. 8.33 The needle pass (blue line) should be perpendicular to the wound margin, so there is no lateral shift of the approximated wound edges

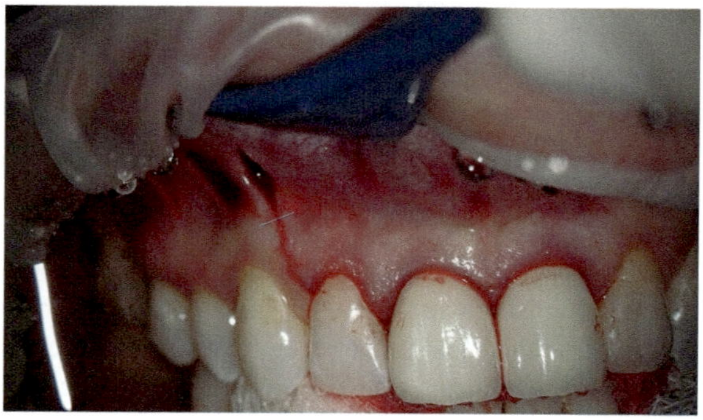

Suture frequency relates to the parallel distance at the tissue surface between two interrupted suture lines. A shorter distance means a higher frequency. This has to be considered because the suture plane has a compression vector, with its maximum force in the suture plane. There are adjacent parallel compression vectors that diminish laterally, which are of equal magnitude to the tissue tension (opening) vectors. Because they dissipate equally on both sides, the adjacent suture should be planned so there is minimal overlap of compression vectors. A rule of thumb to follow is that the distance between two sutures along the wound margin should be equal to the distance between suture entry and exit point on the tissue surface. If sutures are spaced further apart, there will be areas between sutures that are inappropriately approximated. If they are spaced too close together, the opposing tension (opening) vectors adjacent to each suture will overlap, concentrate, and lead to the tissue being less capable to resist tension, in addition to further compromising blood supply (Fig. 8.34).

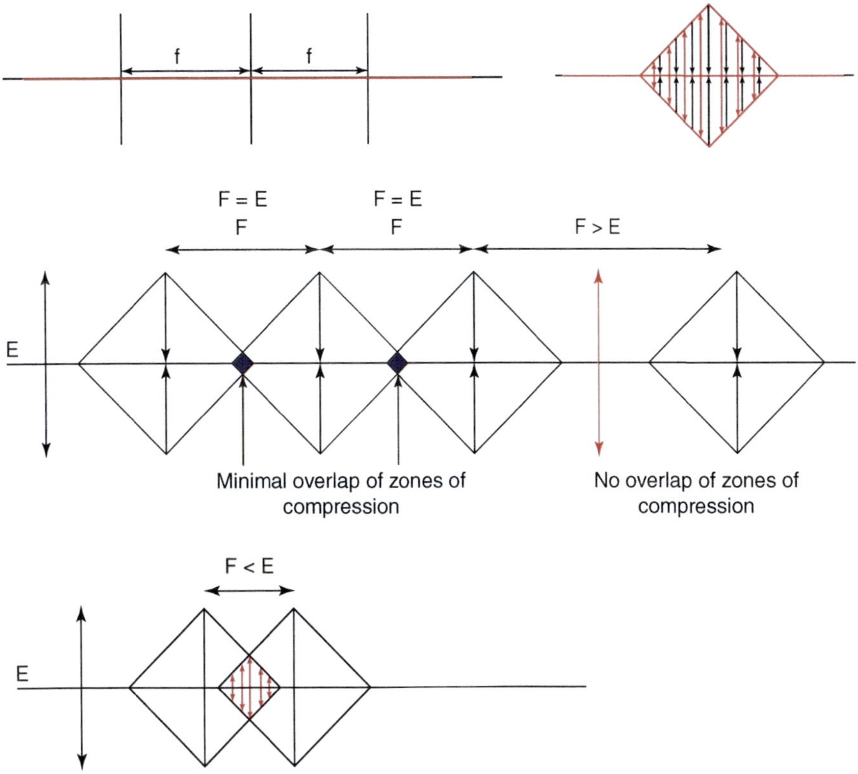

Fig. 8.34 (i) Suture frequency (ii) suture compression vectors perpendicular to wound margin diminish laterally away from suture plane (black arrows) which have equal and opposite tissue tension (opening) vectors (red arrows) (iii) ideal suture frequency is approximately equal to the distance between suture entry and exit points, resulting in minimal overlap of zones of compression (blue shaded areas), (iv) sutures placed too close together concentrate the opening vectors at the wound margin

By knowing all of these aspects of ideal suture geometry, you will then know exactly where and how to precisely place sutures with the microscope. The dimensions (thickness) of the tissue itself will determine suture placement.

Clinical Suturing Steps

The team suturing process begins with checking that the flap is passively positioned—where moving the lip to simulate functional movement (similar to border moulding in denture construction) ensures there is no active displacement. Tissue compression with damp gauze—passed by the second assistant—for a few minutes prior to suturing, helps with approximation and minimising clot thickness. Reasons for the flap sitting in a non-passive situation include: prolonged surgical time, leading to flap shrinkage and dehydration, and insufficient periosteal release under the flap. A medium magnification is used to ensure precise needle placement.

The first assistant starts by slightly retracting the lip and ensuring that the arm is comfortably supported by the armrests. This is to provide access for the operator without displacing the flap. The left hand would be used for mandibular teeth and the right hand for maxillary teeth. An optional step, which may vary among operators, is to use a plastic cheek retractor called the OptraGate (Fig. 8.35). Soft tissues

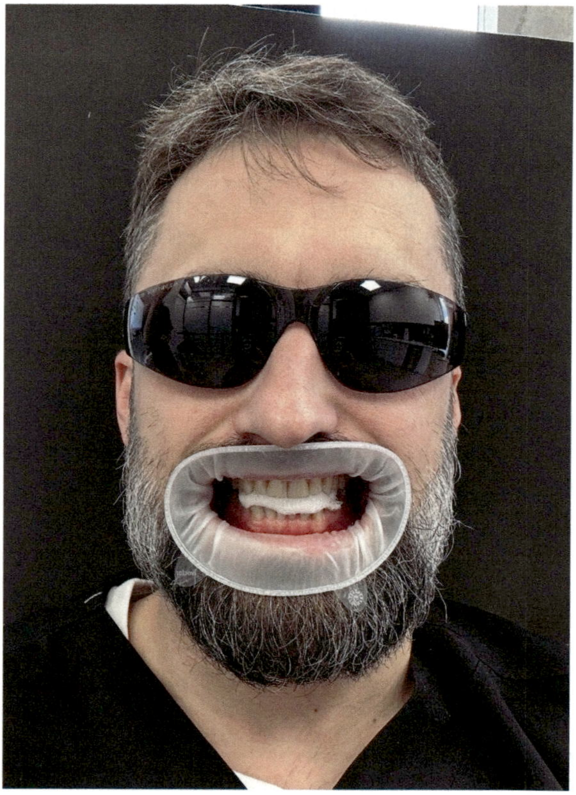

Fig. 8.35 OptraGate® used to retract lips and facial hair

should be dried before it is placed, otherwise the OptraGate will slip. Sometimes this may not be stable if there is a limited vestibular height, or a shallow functional sulcus. For patients with facial hair and beards, this can help provide greater facial hair reflection.

The second assistant places the suture in the needle holder, close to the tip of the beaks, in the body of the needle about one third from the swage—so the suture attachment does not get crushed. There are two directions—needle facing 12 o'clock (Fig. 8.36) and 6 o'clock.

The needle holder, with suture securely held in the jaws, is passed to the operator's right hand in a pen grip. Tissue tweezers for tying are also passed from the second assistant's right hand to the operator's left hand. The plan for the number and order of sutures should be devised beforehand and revised mentally, during flap compression. As a general rule, vertical relieving incisions are sutured first, prior to horizontal submarginal or sulcular incisions. This helps stabilise the flap. An alternative method is to start with horizontal mattress sutures to not only approximate the flap but also to make instrument tying easier. Because the depth of the attached gingiva is known, the bite size is already determined.

Traditionally, the direction of the needle pass has always been taught to be "from free to fixed". This legacy teaching before microscope use was adopted and utilised much larger suture needles. It also was transferred from dentoalveolar surgery which may include suturing mucosa without underlying bone, such as in extraction sockets. In endodontic microsurgery, the suture needles are very fine and therefore not resistant to deformation. In the fixed portion of the flap, accurately passing the needle from the base of the wound surface to the tissue surface, whilst following the geometrical principles described above, will be difficult to achieve. This is because a component of the rotational movement will force the needle point into the bone.

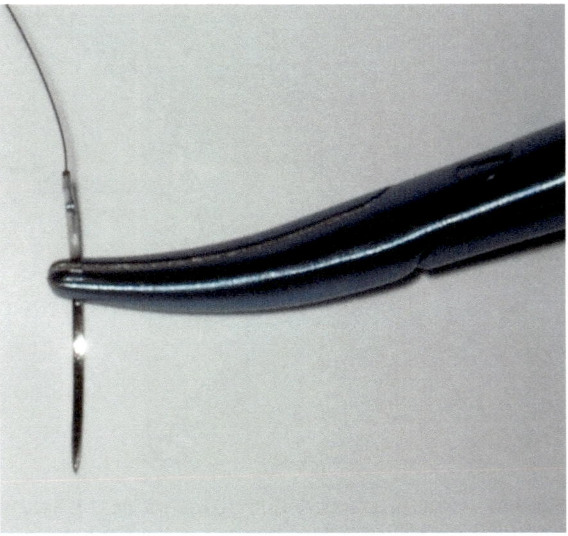

Fig. 8.36 Needle holder with suture facing 12 o'clock

This hard surface prevents the curvature of the needle to precisely enter the wound surface at 90 degrees and exit the tissue at an equidistant position, perpendicular to the tissue surface. By putting pressure on the needle against the bone, it will bend and distort the needle architecture to the point where it can cause tissue damage if used again.

To overcome these limitations, using the scalpel to slightly relieve the underside of the fixed part of the flap adjacent to the wound margin provides room for the suture. This way, the needle can enter at 90 degrees to the wound surface and be precisely guided without resistance to exit perpendicular to the tissue surface.

The operator holds the needle holder in a pen grip, and as the needle passes through the tissue surface, the needle holder is slightly rotated in the fingers. This is why a needle holder with a circular cross section is used, so that the needle holder is rotated in the same axis with a finger and not a wrist movement. The tissue tweezers in the left hand are used to stabilise the part of the flap—this can be either on the free or fixed parts. In most cases, the chord length is long enough so that the needle can enter through the tissue surface, pass through both edges of the wound surface, and exit the tissue surface on the opposite part of the flap—all in one motion. The tissue tweezers are lightly held on the tissue surface, close to the exit point but on the side opposite to the wound margin. This counter pressure provides a reference and a "stop" to guide the needle to the correct exit point. At this point, the needle pass angle can be checked so that it is perpendicular to the wound margin. If not, the suture is removed and the process repeated so the correct needle angle is obtained (Fig. 8.37).

The tissue tweezers in the left hand are then used to hold the needle in the body and pull the rest of the needle and the suture through. The needle holder should not be used to pull the needle from the tip. This is because the closing force can damage the delicate tip and render the needle useless, not to mention it is inefficient, given

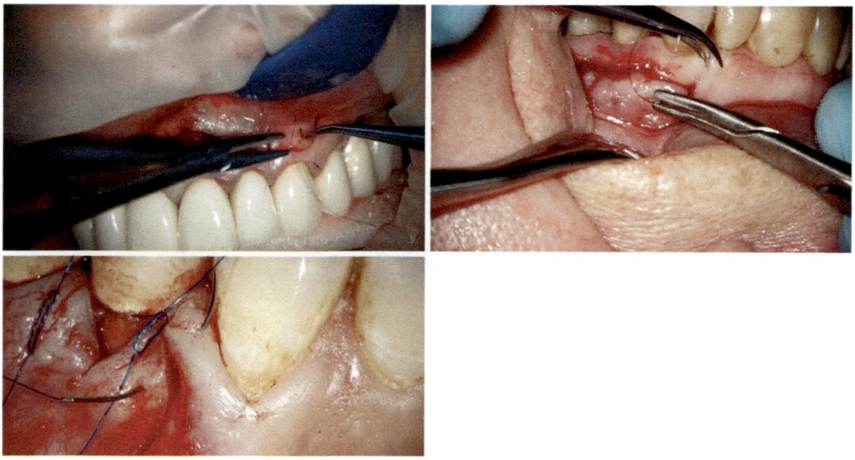

Fig. 8.37 Tissue tweezers stabilising (i) free part and (ii) fixed part of flap during needle pass, (iii) needle showing sufficient chord length to enter and exit tissue surface in one motion

that the left hand is ready and available with the tissue tweezers. If the chord length is not long enough for the needle to make the entire pass in one motion, each half of the wound can have the needle passed through separately. However, the needle should be carefully selected beforehand to have adequate chord length, so this does not occur. The needle holder should be lightly held on the tissue, on the wound margin side of the exit point, to provide a counter pressure when the tissue tweezers pull the suture through. The same hand should be used to pull the suture through in small increments using a class 2 motion. Trying to pull the suture with one continuous movement at the shoulder (class 4) should be avoided. The needle holder is then exchanged with the second assistant with a second pair of tissue tweezers with the curved tips pointing downwards (Fig. 8.38).

Choosing the type of suture knot should be driven by its ideal function (efficacy), and the ability to successfully implement the technique in practice (effectiveness). To ensure precision, tying a 7-0 suture with the microscope should be done using tweezers and not fingers—a technique called an instrument tie because two instruments are used to tie the knot. This ensures only class 1 and 2 movements are used. A technique formerly used was to tie the suture with the fingers of one hand and a needle holder held in the opposite hand. This is not taught in endodontic microsurgery because by doing this, greater force is placed on the suture, leading to risk of tissue strangulation. Physiological tremor is absorbed by the instrument, meaning two instruments tying the knot leads to reduced tremor and greater precision. The instrument tie with more delicate force allows tissues to be approximated, not strangulated. The ideal knot for interrupted sutures in an endodontic microsurgical setting is the 2-2 square (reef) knot. This is a double overhand knot in one direction, followed by a double overhand knot in the other direction (2-2 refers to the number of throws each way). A slight space is left open between the two throws to reduce the increase in pressure caused by the swelling of tissues during the inflammatory response. It also helps prevent possible tissue ischemia that can occur with suture overtightening or strangulation (Fig. 8.39).

This starts by pulling the suture through so that the tail (end) has approximately 40 mm showing from the entry point. Any length longer than this will result in excess wasted movement in tying the first knot and/or more wastage in cutting the tail. A shorter length will have insufficient material to tie the knots. Both hands are using the pen grip to hold a pair of tissue tweezers each. The curved ends of the

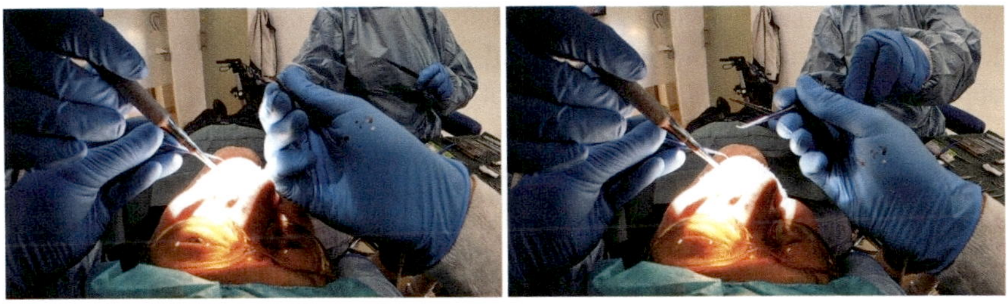

Fig. 8.38 Exchanging needle holder with tissue tweezers

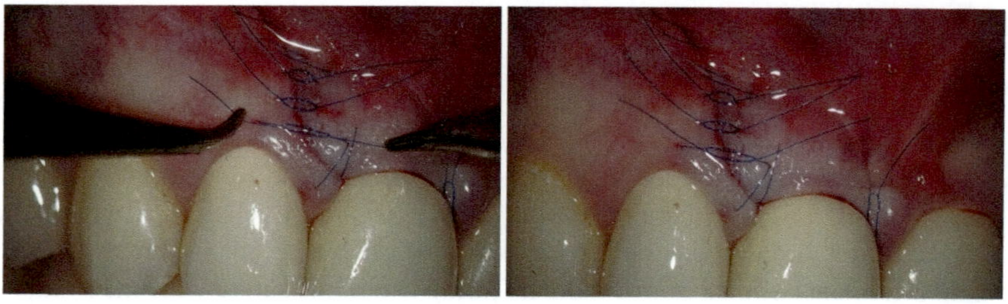

Fig. 8.39 Instrument tie and completed sutures

tweezer tips in both left and right hands should be pointed down, towards the tissue surface. This ensures a more stable and relaxed position for both hands with better access to the suture.

When tying the knot, the hand picking up the tail should start and remain in the position next to the tail. These tweezer tips should be only millimetres from the tail, so it can be "picked" with only a closing motion—typically in the right hand for a right handed operator. This is so there is no unnecessary pulling of the suture or wasted movement. The opposite hand will grasp the needle side of the suture, approximately 80 mm from the exit point. This allows for more room for winding and slack in the suture, without risk of pulling the suture completely through the wound. The "winding" hand brings the suture over to the "picking" tweezers and winds two complete loops over the picking tweezers. The suture tail is then picked at the tip with the "picking" hand. This is so no excess material gets caught in the knot. At this stage, it is important to control the two ends of the tweezers by pulling down towards the tissue and parallel to the needle pass, in a direction towards the opposite side of the wound margin. Meaning, if the right hand is the "picking" hand which picks the tail on the coronal side of a horizontal submarginal incision, after winding two loops and securing the tail, the right hand ties the knot by pulling it towards the apical side of the incision (i.e. opposite to the coronal side). Both edges of the wound should be carefully approximated but not strangulated, which would compress and damage the tissue. This first loop with two throws now brings the tail to the opposite side compared to prior to making the first loop. For the second loop, this process is repeated again, again with the picking tweezers held right next to the tail. The second loop is wound in the opposite direction to the first loop and again, both ends of the suture are pulled parallel to the needle pass, perpendicular to the wound margin and opposite to the previous tying direction, in order to close the loop. It is important to tighten the second loop with precision, so that it does not approximate with the first loop. A space of approximately 2 mm in diameter is left between loops, which allows for tissue swelling from inflammation. This design tightens the knot with swelling because of the space between loops, preventing strangulation during the healing process. To cut the suture, the second assistant exchanges the tissue tweezers with the micro scissors in the operator's right hand. For stability and to aid in stereopsis, the side of the scissors is held in contact with the tweezers in the left hand during the cutting process. Approximately 20 mm of the suture is left on each side (Fig. 8.40).

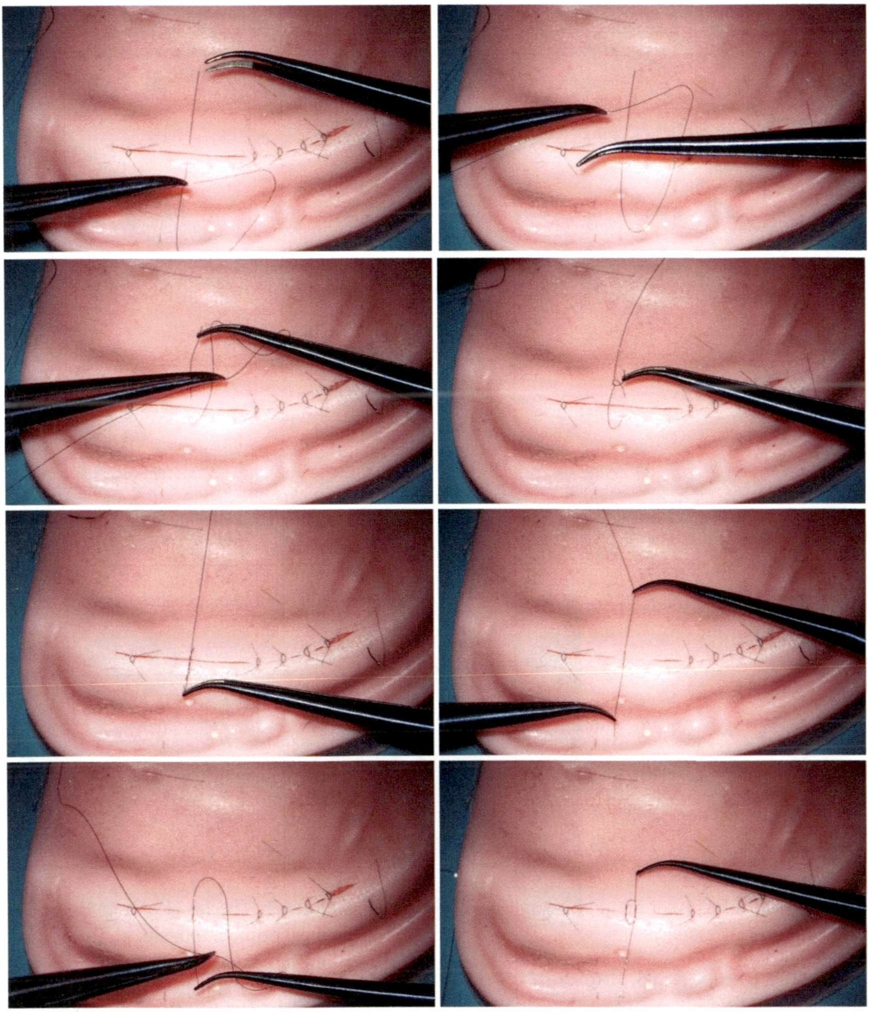

Fig. 8.40 (i) Correct position for picking tweezers to be held (right next to tail), (ii) incorrect position (further away from tail), (iii) winding tweezer has performed two overhand throws over picking tweezers from bottom of image towards top and picking tweezer picks suture very close to the suture tail, (iv) winding tweezer pulls suture over picking tweezer, (v) picking tweezer pulls tail towards opposite side, (vi) both tweezers hold suture closer to the knot and are pulling parallel to the needle pass in opposite directions to close the knot with precision, (vii) second knot is tied with two overhand throws over picking tweezers in opposite direction—from top of image towards bottom, (viii) tail is pulled in opposite direction, (ix) tweezers held close to the knot to close with precision, (x and xi) scissors touching tweezers to aid in stereopsis when cutting both suture tails

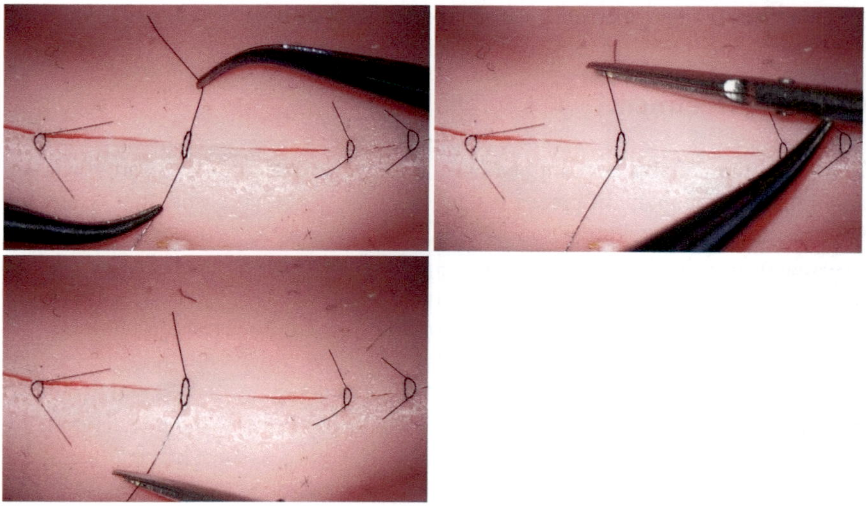

Fig. 8.40 (continued)

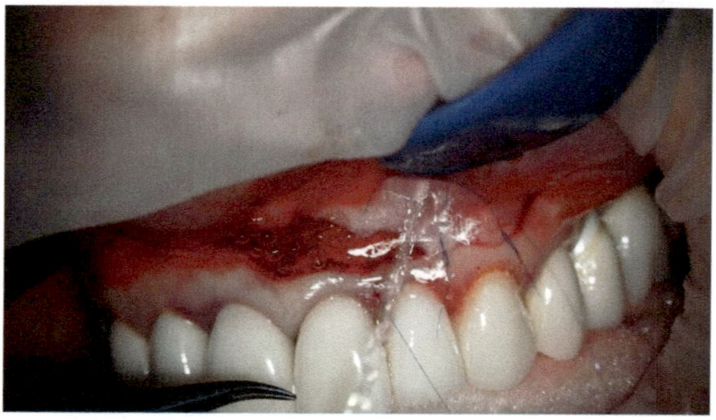

Fig. 8.41 First assistant wetting flap during suturing

During this process, the second assistant picks up the needle and places it in the needle holder, facing either 12 o'clock or 6 o'clock, ready for the next suture to be placed. It is important for the operator to be delicate when handling the suture. If it is held too firm with the tweezers, the strand can get twisted or have bends/kinks placed in it, which not only can affect the tissue as it is pulled through, but lead to inefficient instrument ties. The flap can also be hydrated during suturing by the first assistant (Fig. 8.41).

Video link to suturing using the microscope:

These sutures described above are for precision approximation along the incision line. Depending on the flap design, stabilising sutures may be placed first which maintain the flap in an ideal position, prior to approximation. Another method prior to suturing the wound margin is flap compression with damp gauze, which not only mechanically stabilises the flap but hydrates it, minimising contraction that occurs with dehydration during extended surgery duration. Blood clot thickness under the flap is also reduced which aids in healing. The ideal method for stabilising a submarginal flap design is by using a horizontal mattress suture, which keeps the knot away from the wound margin and wound surface. The horizontal mattress design ties the knot parallel to the incision line. Because of the greater strength required, a larger 6-0 suture is used (Fig. 8.42).

Tissue tweezers are passed to the operator's left hand by the second assistant. The suture is placed in the needle holder by the second assistant (facing 6 o'clock for an upper anterior tooth) and passed to the operator's right hand. Using the tissue tweezers, the free part of the flap is held lightly and the needle passed through the centre of the flap, where the knot will be tied. The needle will then be completely pulled through and placed in the needle holder again. This time it is passed under the wound surface on the fixed part of the flap to exit in the attached gingiva. Because the needle is larger than 7-0, it is strong enough to resist deformation from curving the needle against bone. Once it is pulled out of the exit point using the tissue tweezers, the needle is placed in the needle holder, now facing 12 o'clock for an upper anterior tooth. The entry point in the attached gingiva is along an imaginary line going through the exit point, parallel to the incision line. This time, the needle pass through the attached gingiva goes towards the free part of the flap and perpendicular to the incision line (i.e. parallel to the first needle pass). The needle is pulled through with the tissue tweezers and placed again in the needle holder in the right hand. Using the tissue tweezers, the free part of the flap is held in the left hand and

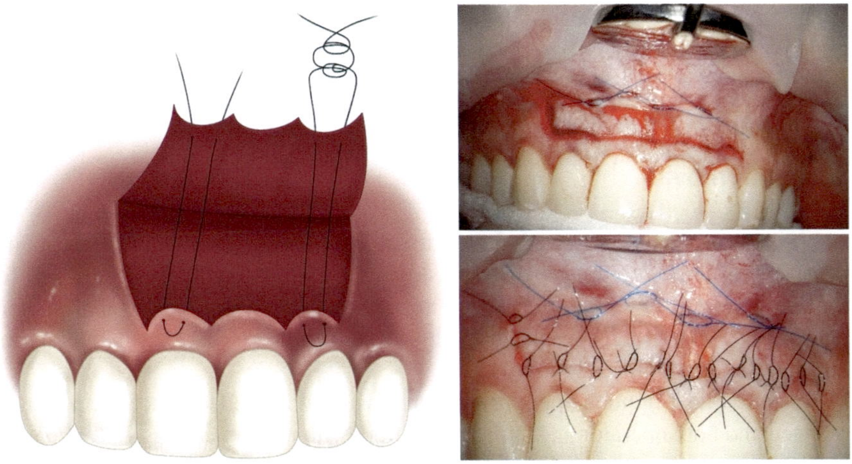

Fig. 8.42 (i) Horizontal mattress suture diagram, (ii) initial placement to help stabilise flap, (iii) completed sutures

the needle enters at a point on a line going through the exit point and parallel to the incision line—at a distance equal to the entry and exit points in the attached gingiva. The needle holder is exchanged with the second tissue tweezers for tying the knot. The 2-2 square knot for the horizontal mattress suture is tied in the same way as mentioned above in the single interrupted suture, except the strands are pulled parallel to the incision line. The aim is to keep the flap stabilised, not closely approximated, meaning the horizontal mattress suture is loosely tied and accounts for post-operative swelling in the flap to tighten the knot and prevent strangulation.

After the Appointment

Once all the sutures are placed, flap compression with damp gauze for 15 min is done again to provide haemostasis, hydration, and flap stability of the surgical site. It helps approximate the flap and minimise clot thickness. To check for a passive flap, the lip is lifted and moved in all directions to check for movement of the flap and along the incision line. If there is tension or movement, the sutures may need to be adjusted, more placed, or more flap compression required. If all steps regarding tissue compression and suture placement have been followed, any further modification would be unlikely.

In some cases, post-operative LA can be administered for haemostasis and pain relief, particularly if the procedure is longer than usual. Post-operative instructions are then given, which are to ensure the patient understands how to not disturb the flap, not elicit bleeding and keep the blood clot present to provide the greatest chance of healing. If cone beam-computed tomography is available on site, a post-operative CBCT scan is recommended not only to have a comparison when the review CBCT is taken, but also to get feedback on the dimensions of the root end preparation. In some countries, this may be a controversial step regarding radiation exposure; however, current small field of view CBCT machines have a very low effective radiation dose. There is also a benefit to the patient, especially if a comparative decision regarding the amount of healing needs to be made in the short term.

Suture Removal

This visit is usually scheduled 2–3 days following the surgical appointment, mainly because the sutures do not provide any strength benefit beyond this time. If left for longer, there may be tissue overgrowth impeding ease of suture removal. The main factors to check for are: patient symptoms, oral hygiene, primary healing along the incision line and suture integrity.

Because this is a quick appointment, one assistant is needed. Once the patient is reclined and both operator and assistant are in position, the microscissors are passed to the operator's right hand. The side of the knot is cut in an area furthest away from incision line, which is very close to the exit point on the tissue surface. The microscissors are then exchanged with the tweezers which grasp the suture in the middle of the knot. This ensures the portion exposed to the oral cavity is removed without being pulled through the suture track. If there is some tissue overgrowth, an explorer can be used to "lift" the knot so the microscissors can have access to cut the suture. All of this is done with a medium magnification, so that the suture can be removed with minimal disturbance to the soft tissue. Depending on the sutures to be removed, it may be more efficient for the operator to cut all sutures, exchange the microscissors with tweezers, then remove all the sutures with gauze in the assistant's left hand used to collect each suture.

Training

The concepts with microsurgery training are the same as those for non-surgical procedures. The entire procedure is rehearsed without a patient, every step is repeated with feedback until it is performed perfectly and the simulation with all clinical steps is done, so the correct patterns can be stored by operator and both assistants.

Video recording is also used to each team member can review their own actions and visualise how best to improve them. To help adapt to both roles, assistants should train in both the first and second position.

Microsurgery Steps

Chlorhexidine rinse
LA
Rehearse patient head movements
Gauze x2
Pass retractor (1st) and scalpel (2nd)
2nd: suction and Stropko
Incision
Elevation (medium magnification)
Osteotomy (low magnification)
2nd: two suctions
Root resection (low, then medium magnification)
Curettage
Methylene blue, wash
Micro mirror (high magnification)
Adrenaline pellet
Ultrasonic preparation (low, then high magnification)
Filling (low, then high magnification)
Moist cotton pellet
Remove adrenaline pellet
Check PA radiograph
Damp gauze for compression (2nd)

Retraction (1st)
Suturing (medium magnification, low if horizontal mattress suture is placed)
Damp gauze for compression (2nd)
Check flap movement
Post-operative LA (optional)
Post-operative instructions
CBCT scan

References

Burkhardt R, Lang NP. Coverage of localized gingival recessions: comparison of micro- and macrosurgical techniques. J Clin Periodontol. 2005;32(3):287–93.

Burkhardt R, Preiss A, Joss A, Lang NP. Influence of suture tension to the tearing characteristics of the soft tissues: an in vitro experiment. Clin Oral Implants Res. 2008;19(3):314–9.

Chan H-L, Velasquez-Plata D. Microsurgery in periodontal and implant dentistry concepts and applications. 1st ed. Cham: Springer; 2022.

Hirata T. The influence of microsurgical protocols on wound healing study in vivo. Int J Microdent. 2012;3(1):15–9.

Konz SA, Johnson S. Work design: occupational ergonomics. 7th ed. Scottsdale, AZ: Holcomb Hathaway; 2008.

Macsai MS, editor. Ophthalmic microsurgical suturing techniques. 1st ed. Berlin: Springer; 2007.

Shanelec DA. Periodontal microsurgery. J Esthet Restor Dent. 2003;15(7):402–7. discussion 408

Correction to: Team Microscope Endodontics

Kehn E. Yapp and Carlos Aznar Portoles

Correction to:
Chapters 3 and 8 in: K. E. Yapp, C. Aznar Portoles,
Team Microscope Endodontics,
https://doi.org/10.1007/978-3-031-70188-7

Contributors' names and affiliations have been included in the chapter opening pages of the chapters 3 and 8 and the corrected version has been uploaded.

The updated version of these chapters can be found at
https://doi.org/10.1007/978-3-031-70188-7_3
https://doi.org/10.1007/978-3-031-70188-7_8

K. E. Yapp, C. Aznar Portoles, *Team Microscope Endodontics*,
https://doi.org/10.1007/978-3-031-70188-7_9

Closing Remarks

After reading this book, you should now have a complete understanding of how to set up a microscope-centred office, train your team, and perform all endodontic procedures using the microscope. By focusing on these details, it easily increases your ability to do all types of dental treatment using the microscope. These concepts can be applied to restorative, periodontal, and preventive dentistry.

We encourage you to provide regular training meetings with your team, especially when new staff members join. This way you can constantly refresh your teamwork skills and ensure consistency with all operators and assistants.

If you are looking to open a new office, it is important you have mentoring from someone who has completed a microscope-centred new office setup. For access to mentoring and feedback from the authors, you can contact us on support@quantumergonomics.com.

Once your team microscope journey is complete, we hope your practice and personal life flourishes. The rewards are endless and we would like to hear of stories about how patient management has improved, team efficiency has been embraced, and your physical health has been maintained to a high level.

We are honoured to have the opportunity to write this book. It is our hope that the next generation will be able to embrace these team microscope skills and raise the level of patient care with the highest standard of expertise.

Kehn E. Yapp
Carlos Aznar Portoles

K. E. Yapp, C. Aznar Portoles, *Team Microscope Endodontics*,
https://doi.org/10.1007/978-3-031-70188-7

GPSR Compliance

The European Union's (EU) General Product Safety Regulation (GPSR) is a set of rules that requires consumer products to be safe and our obligations to ensure this.

If you have any concerns about our products, you can contact us on ProductSafety@springernature.com

In case Publisher is established outside the EU, the EU authorized representative is:

Springer Nature Customer Service Center GmbH
Europaplatz 3
69115 Heidelberg, Germany

Batch number: 10641516

Printed by Printforce, the Netherlands